W9-BCI-620

REA's Books Are The Best...
They have rescued lots of grades and more!

(a sample of the <u>hundreds of letters</u> REA receives each year)

"Your books are great! They are very helpful, and have upped my grade in every class. Thank you for such a great product."

Student, Seattle, WA

"Your book has really helped me sharpen my skills and improve my weak areas. Definitely will buy more."

Student, Buffalo, NY

"Compared to the other books that my fellow students had, your book was the most useful in helping me get a great score."

Student, North Hollywood, CA

"I really appreciate the help from your excellent book. Please keep up your great work."

Student, Albuquerque, NM

"Your book was such a better value and was so much more complete than anything your competition has produced (and I have them all)!"

Teacher, Virginia Beach, VA

(more on next page)

(continued from previous page)

" Your books have saved my GPA, and quite possibly my sanity.
My course grade is now an 'A', and I couldn't be happier. "

Student, Winchester, IN

" These books are the best review books on the market.
They are fantastic! "

Student, New Orleans, LA

" Your book was responsible for my success on the exam. . . I
will look for REA the next time I need help. "

Student, Chesterfield, MO

" I think it is the greatest study guide I have ever used! "

Student, Anchorage, AK

" I encourage others to buy REA because of their superiority.
Please continue to produce the best quality books on the market. "

Student, San Jose, CA

" Just a short note to say thanks for the great support your book
gave me in helping me pass the test . . . I'm on my way to a
B.S. degree because of you ! "

Student, Orlando, FL

Super Review™

All You Need to Know!

LINEAR ALGEBRA

**By the Staff of
Research & Education Association
Dr. M. Fogiel, Director**

Research & Education Association
61 Ethel Road West
Piscataway, New Jersey 08854

SUPER REVIEW ™
OF LINEAR ALGEBRA

Printed in the United States of America

Library of Congress Catalog Card Number 00-130289

International Standard Book Number 0-87891-085-9

SUPER REVIEW is a trademark of
Research & Education Association, Piscataway, New Jersey 08854

WHAT THIS Super Review WILL DO FOR YOU

This **Super Review** provides all that you need to know to do your homework effectively and succeed on exams and quizzes.

The book focuses on the core aspects of the subject, and helps you to grasp the important elements quickly and easily.

Outstanding **Super Review** features:

- Topics are covered in logical sequence

- Topics are reviewed in a concise and comprehensive manner

- The material is presented in student-friendly language that makes it easy to follow and understand

- Individual topics can be easily located

- Provides excellent preparation for midterms, finals and in-between quizzes

- In every chapter, reviews of individual topics are accompanied by Questions **Q** and Answers **A** that show how to work out specific problems

- At the end of most chapters, quizzes with answers are included to enable you to practice and test yourself to pinpoint your strengths and weaknesses

- Written by professionals and test experts who function as your very own tutors

Dr. Max Fogiel
Program Director

CONTENTS

4   LINEAR TRANSFORMATIONS

5   EIGENVALUES AND EIGENVECTORS

CHAPTER 1

Linear Matrices

1.1 Linear Equations and Matrices

A linear equation is an equation of the form $a_1 x_1 + a_2 x_2 + \ldots + a_n x_n = b$, where $a_1, \ldots, a_n$ and b are real constants.

EXAMPLES

a) $2x + 6y = 9$

b) $x_1 + 3x_2 + 7x_3 = 5$

c) $\alpha - 2 = 0$

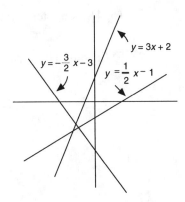

Figure 1.1 Linear equations in two variables are always straight lines.

A system of linear equations is a finite set of linear equations, all of which use the same set of variables.

EXAMPLES

a) $2x_1 + x_2 + 5x_3 = 4$
$x_2 + 3x_3 = 0$
$7x_1 + 3x_2 + x_3 = 9$

b) $y - z = 5$
$z = 1$

The solution of a system of linear equations is that set of real numbers which, when substituted into the set of variables, satisfies each equation in the system. The set of all solutions is called the solution set S of the system.

EXAMPLE

$y + z = 9 \qquad S = \{5, 4\}$
$z = 4$

A consistent system of linear equations has at least one solution, while an inconsistent system has no solutions.

EXAMPLES

a) $y + z = 9 \qquad S = \{5, 4\}$ (consistent system)
$z = 4$

b) $x_1 + x_2 = 7 \qquad S = \varnothing$ (inconsistent system)
$x_1 = 3$
$x_1 - x_2 = 7$

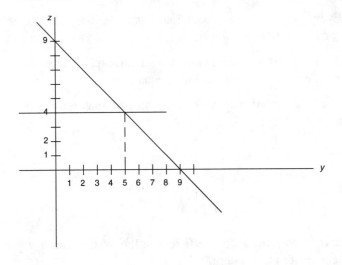

Figure 1.2 Consistent System

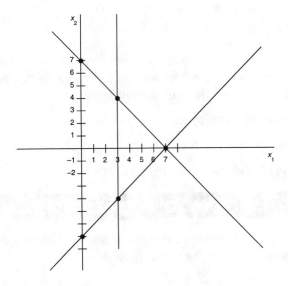

Figure 1.3 Inconsistent System

Every system of linear equations has either one solution, no solution, or infinitely many solutions.

A system of linear equations with infinitely many solutions is called a dependent system of linear equations.

The augmented matrix for a system of linear equations is the matrix of the form:

$$\left[\begin{array}{cccc|c} a_{11} & a_{12} \cdots & a_{1n} & & b_1 \\ a_{21} & a_{22} \cdots & a_{2n} & & b_2 \\ \vdots & & & & \\ a_{m1} & a_{m2} \cdots & a_{mn} & & b_m \end{array}\right]$$

where a_{ij} represents each coefficient in the system and b_i represents each constant in the system.

EXAMPLE

$$\begin{aligned} x_1 + 6x_2 - 2x_3 &= 4 \\ 3x_1 + x_3 &= 7 \\ 5x_1 - 3x_2 + x_3 &= 0 \end{aligned} \qquad \left[\begin{array}{ccc|c} 1 & 6 & -2 & 4 \\ 3 & 0 & 1 & 7 \\ 5 & -3 & 1 & 0 \end{array}\right]$$

Elementary row operations are operations on the rows of an augmented matrix, which are used to reduce that matrix to a more solvable form. These operations are the following:

a) Multiply a row by a non-zero constant.

b) Interchange two rows.

c) Add a multiple of one row to another row.

Problem Solving Examples:

By forming the augmented matrix and row reducing, determine the solutions of the following system:

$$\begin{aligned} 2x - y + 3z &= 4 \\ 3x + 2z &= 5 \\ -2x + y + 4z &= 6 \end{aligned}$$

The augmented matrix of the system is:

$$\begin{bmatrix} 2 & -1 & 3 & | & 4 \\ 3 & 0 & 2 & | & 5 \\ -2 & 1 & 4 & | & 6 \end{bmatrix}.$$

Add the first row to the third row:

$$\begin{bmatrix} 2 & -1 & 3 & | & 4 \\ 3 & 0 & 2 & | & 5 \\ 0 & 0 & 7 & | & 10 \end{bmatrix}$$

This is the augmented matrix of:

$$\begin{aligned} 2x - y + 3z &= 4 \\ 3x \phantom{{}-y} + 2z &= 5 \\ 7z &= 10 \end{aligned}$$

The system has been sufficiently simplified now so that the solution can be found.

From the last equation we have $z = {}^{10}\!/_{7}$. Substituting this value into the second equation and solving for x gives $x = {}^{5}\!/_{7}$. Substituting $x = {}^{5}\!/_{7}$ and $z = {}^{10}\!/_{7}$ into the first equation and solving for y yields $y = {}^{12}\!/_{7}$. The solution to the system is, therefore,

$$x = \frac{5}{7}, \; y = \frac{12}{7}, \; z = \frac{10}{7}.$$

 Solve the following linear system of equations:

$$\begin{aligned} 2x + 3y - 4z &= 5 \\ -2x \phantom{{}+ 3y} + z &= 7 \\ 3x + 2y + 2z &= 3 \end{aligned}$$

 The augmented matrix for the system is:

$$\begin{bmatrix} 2 & 3 & -4 & | & 5 \\ -2 & 0 & 1 & | & 7 \\ 3 & 2 & 2 & | & 3 \end{bmatrix}$$

which can be reduced by using the following sequence of row operations:

Add the first row to the second row.

$$\begin{bmatrix} 2 & 3 & -4 & | & 5 \\ 0 & 3 & -3 & | & 12 \\ 3 & 2 & 2 & | & 3 \end{bmatrix}$$

Divide the first row by 2 and the second row by 3.

$$\begin{bmatrix} 1 & \frac{3}{2} & -2 & | & \frac{5}{2} \\ 0 & 1 & -1 & | & 4 \\ 3 & 2 & 2 & | & 3 \end{bmatrix}$$

Add -3 times the first row to the third row.

$$\begin{bmatrix} 1 & \frac{3}{2} & -2 & | & \frac{5}{2} \\ 0 & 1 & -1 & | & 4 \\ 0 & -\frac{5}{2} & 8 & | & -\frac{9}{2} \end{bmatrix}$$

Add $\frac{5}{2}$ times the second row to the third row.

$$\begin{bmatrix} 1 & \frac{3}{2} & -2 & | & \frac{5}{2} \\ 0 & 1 & -1 & | & 4 \\ 0 & 0 & \frac{11}{2} & | & \frac{11}{2} \end{bmatrix}$$

This is the augmented matrix for the system:

$$x + \tfrac{3}{2}y - 2z = \tfrac{5}{2}$$
$$y - z = 4$$
$$\tfrac{11}{2}z = \tfrac{11}{2}$$

Now the solution to this system can be easily found. From the last equation we have $z = 1$. Substituting $z = 1$ in the second equation gives $y = 5$. Next, substitute $y = 5$ and $z = 1$ into the first equation. This gives $x = -3$. Therefore, the solution to the system is $x = -3$, $y = 5$, $z = 1$.

Solve the following system:

$$x + y + 2z = 9$$
$$2x + 4y - 3z = 1$$
$$3x + 6y - 5z = 0$$

 The augmented matrix for the system is:

$$\begin{bmatrix} 1 & 1 & 2 & 9 \\ 2 & 4 & -3 & 1 \\ 3 & 6 & -5 & 0 \end{bmatrix}.$$

It can be reduced by elementary row operations.

Add −2 times the first row to the second row and −3 times the first row to the third row.

$$\begin{bmatrix} 1 & 1 & 2 & 9 \\ 0 & 2 & -7 & -17 \\ 0 & 3 & -11 & -27 \end{bmatrix}$$

Multiply the second row by $1/2$.

$$\begin{bmatrix} 1 & 1 & 2 & 9 \\ 0 & 1 & -\frac{7}{2} & -\frac{17}{2} \\ 0 & 3 & -11 & -27 \end{bmatrix}$$

Add −3 times the second row to the third row.

$$\begin{bmatrix} 1 & 1 & 2 & 9 \\ 0 & 1 & -\frac{7}{2} & -\frac{17}{2} \\ 0 & 0 & -\frac{1}{2} & -\frac{3}{2} \end{bmatrix}$$

Multiply the third row by −2 to obtain

$$\begin{bmatrix} 1 & 1 & 2 & 9 \\ 0 & 1 & -\frac{7}{2} & -\frac{17}{2} \\ 0 & 0 & 1 & 3 \end{bmatrix}.$$

This is the augmented matrix for the system:

$$x + y + 2z = 9$$
$$y - \frac{7}{2}z = -\frac{17}{2}$$
$$z = 3$$

Solving this system gives $x = 1$, $y = 2$, and $z = 3$.

 For the following system, find the augmented matrix; then, by reducing, determine whether the system has a solution.

$$3x - y + z = 1$$
$$7x + y - z = 6 \tag{1}$$
$$2x + y - z = 2$$

 The augmented matrix for the system is

$$\begin{bmatrix} 3 & -1 & 1 & 1 \\ 7 & 1 & -1 & 6 \\ 2 & 1 & -1 & 2 \end{bmatrix}.$$

This can be reduced by performing the following row operations. Divide the first row by 3.

$$\begin{bmatrix} 1 & -\frac{1}{3} & \frac{1}{3} & \frac{1}{3} \\ 7 & 1 & -1 & 6 \\ 2 & 1 & -1 & 2 \end{bmatrix}$$

Now add –7 times the first row to the second row and –2 times the first row to the third row.

$$\begin{bmatrix} 1 & -\frac{1}{3} & \frac{1}{3} & \frac{1}{3} \\ 0 & \frac{10}{3} & -\frac{10}{3} & \frac{11}{3} \\ 0 & \frac{5}{3} & -\frac{5}{3} & \frac{4}{3} \end{bmatrix}$$

Divide the second row by $^{10}\!/_3$, and add $-^5\!/_3$ times the second row to the third row.

$$\begin{bmatrix} 1 & -\frac{1}{3} & \frac{1}{3} & \Big| & \frac{1}{3} \\ 0 & 1 & -1 & \Big| & \frac{11}{10} \\ 0 & 0 & 0 & \Big| & -\frac{1}{2} \end{bmatrix}$$

This is the augmented matrix of the system:

$$\begin{aligned} x - \tfrac{1}{3}y + \tfrac{1}{3}z &= \tfrac{1}{3} \\ y - z &= \tfrac{11}{10} \\ 0 &= -\tfrac{1}{2} \end{aligned} \qquad (2)$$

The last equation cannot hold for any choice of x, y, and z. Thus, system (2) has no solution. Therefore, system (1) has no solution.

1.2 Homogeneous Systems of Linear Equations

A homogeneous system of linear equations is a system in which all of the constant terms (those which are not multiplied by any variables) are zero.

EXAMPLE

$x_1 + 3x_2 = 0$

$4x_1 + x_2 + 7x_3 = 0$ is a homogeneous system.

$2x_2 + 2x_3 = 0$

Every homogeneous system of n linear equations has at least one solution, called the trivial solution, in which all of the variables x_1, $x_2, \ldots, x_n$ are equal to zero. All other solutions to the system are called non-trivial solutions.

Every homogeneous system of linear equations has either (a) only the trivial solution, or (b) the trivial solution and an infinite number of non-trivial solutions. If there are more unknowns than equations, then the system has non-trivial solutions.

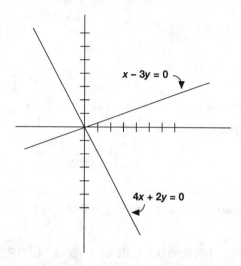

Figure 1.4 In a homogeneous system, all lines pass through the origin.

Problem Solving Examples:

 Solve the following system of equations:

$$x + 3y = 0$$
$$2x + 6y + 4z = 0 \tag{1}$$

To solve the given system of equations, first form the matrix that consists of the coefficients only. Then reduce this matrix. (The augmented matrix is not needed since all of the constant terms are zero.) The matrix of the coefficients is

$$\begin{bmatrix} 1 & 3 & 0 \\ 2 & 6 & 4 \end{bmatrix}.$$

Now add –2 times the first row to the second row.

$$\begin{bmatrix} 1 & 3 & 0 \\ 0 & 0 & 4 \end{bmatrix}$$

Divide the second row by 4.

$$\begin{bmatrix} 1 & 3 & 0 \\ 0 & 0 & 1 \end{bmatrix}$$

The previous is the matrix of coefficients for:

$$x + 3y = 0$$
$$z = 0 \tag{2}$$

This system is easy to solve. We have $z = 0$ and can assign y any value. Then compute x from (2). This gives a solution to (1).

 Solve the following homogeneous system of linear equations.

$$2x_1 + 2x_2 - x_3 + x_5 = 0 \tag{1}$$
$$-x_1 - x_2 + 2x_3 - 3x_4 + x_5 = 0$$
$$x_1 + x_3 - 2x_3 - x_5 = 0$$
$$x_3 + x_4 + x_5 = 0$$

A System (1) has five unknowns but only four equations. We know that a homogeneous system of linear equations with more unknowns than equations has a non-zero (non-trivial) solution. To solve system (1), form the matrix of the coefficients. Then reduce this matrix. The coefficient matrix is:

$$A = \begin{bmatrix} 2 & 2 & -1 & 0 & 1 \\ -1 & -1 & 2 & -3 & 1 \\ 1 & 1 & -2 & 0 & -1 \\ 0 & 0 & 1 & 1 & 1 \end{bmatrix}. \tag{1}$$

Add the fourth row to the first row and the third row to the second row.

$$\begin{bmatrix} 2 & 2 & 0 & 1 & 2 \\ 0 & 0 & 0 & -3 & 0 \\ 1 & 1 & -2 & 0 & -1 \\ 0 & 0 & 1 & 1 & 1 \end{bmatrix} \tag{2}$$

Divide the second row by -3. Then add -1 times the second row to the first row and to the fourth row.

$$\begin{bmatrix} 2 & 2 & 0 & 0 & 2 \\ 0 & 0 & 0 & 1 & 0 \\ 1 & 1 & -2 & 0 & -1 \\ 0 & 0 & 1 & 0 & 1 \end{bmatrix} \tag{3}$$

Divide the first row by 2. Then add -1 times the first row to the third row.

$$\begin{bmatrix} 1 & 1 & 0 & 0 & 1 \\ 0 & 0 & 0 & 1 & 0 \\ 0 & 0 & -2 & 0 & -2 \\ 0 & 0 & 1 & 0 & 1 \end{bmatrix} \tag{4}$$

Add 2 times the fourth row to the third row.

$$\begin{bmatrix} 1 & 1 & 0 & 0 & 1 \\ 0 & 0 & 0 & 1 & 0 \\ 0 & 0 & 0 & 0 & 0 \\ 0 & 0 & 1 & 0 & 1 \end{bmatrix} \tag{5}$$

Interchange the second and fourth rows. Next, interchange the third and fourth rows.

$$\begin{bmatrix} 1 & 1 & 0 & 0 & 1 \\ 0 & 0 & 1 & 0 & 1 \\ 0 & 0 & 0 & 1 & 0 \\ 0 & 0 & 0 & 0 & 0 \end{bmatrix} \tag{6}$$

The corresponding system of equation is:

$$x_1 + x_2 + x_5 = 0$$
$$x_3 + x_5 = 0$$
$$x_4 = 0$$

Solving for the leading variables yields:

$$x_1 = -x_2 - x_5$$
$$x_3 = -x_5$$
$$x_4 = 0$$

The solution set is, therefore, given by:

$$x_1 = -s - t, x_2 = s, x_3 = -t, x_4 = 0, x_5 = t.$$

That is, we have chosen x_2 and x_5 to be free variables and, hence, set them equal to the parameters s and t, respectively. The dependent variables are then x_1 and x_3, and their dependence on s and t are given by the reduced form of the system. Thus, any solution is of the form $(-s - t, s, -t, 0, t)$.

[A] Show that each of the following systems has a non-zero solution:

(a)
$$x + 2y - 3z + w = 0 \qquad (1)$$
$$x - 3y + z - 2w = 0$$
$$2x + y - 3z + 5w = 0$$

(b)
$$x + y - z = 0 \qquad (2)$$
$$2x - 3y + z = 0$$
$$x - 4y + 2z = 0$$

[B] Show that the following system has a unique solution:

$$x + y - z = 0$$
$$2x + 4y - z = 0 \qquad (3)$$
$$3x + 2y + 2z = 0$$

[A] (a) System (1) has a non-zero solution since there are four unknowns but only three equations.

(b) The coefficient matrix for system (2) is

$$\begin{bmatrix} 1 & 1 & -1 \\ 2 & -3 & 1 \\ 1 & -4 & 2 \end{bmatrix}.$$

Add -2 times the first row to the second row and -1 times the first row to the third row.

$$\begin{bmatrix} 1 & 1 & -1 \\ 0 & -5 & 3 \\ 0 & -5 & 3 \end{bmatrix}$$

Add −1 times the second row to the third row.

$$\begin{bmatrix} 1 & 1 & -1 \\ 0 & -5 & 3 \\ 0 & 0 & 0 \end{bmatrix}$$

This is the matrix of coefficients of the system:

$$x + y - z = 0$$
$$-5y + 3z = 0$$
$$0 = 0$$

The system has a non-zero solution since we obtained only two equations in the three unknowns when we reduced the system to echelon form. For example, let $z = 5$; then $y = 3$ and $x = 2$ solves the system.

[B] The matrix of coefficients for system (3) is:

$$\begin{bmatrix} 1 & 1 & -1 \\ 2 & 4 & -1 \\ 3 & 2 & 2 \end{bmatrix}.$$

Add −2 times the first row to the second row and −3 times the first row to the third row.

$$\begin{bmatrix} 1 & 1 & -1 \\ 0 & 2 & 1 \\ 0 & -1 & 5 \end{bmatrix}$$

Multiply the third row by 2. Then, add the second row to the third row.

$$\begin{bmatrix} 1 & 1 & -1 \\ 0 & 2 & 1 \\ 0 & 0 & 11 \end{bmatrix}$$

The corresponding system of equations is:

$$x + y - z = 0$$
$$2y + z = 0$$
$$11z = 0$$

Solving this system yields $x = y = z = 0$. In general, a system of homogeneous equations in n unknowns has a zero solution if the corresponding reduced matrix has exactly n rows with non-zero entries.

 Solve the following system of equations by forming the matrix of coefficients and reducing.

$$3x + 2y - z = 0 \qquad\qquad (1)$$
$$x - y + 2z = 0$$
$$x + y - 6z = 0$$

 The matrix of coefficients of system (1) is:

$$\begin{bmatrix} 3 & 2 & -1 \\ 1 & -1 & 2 \\ 1 & 1 & -6 \end{bmatrix}.$$

Add -3 times the second row to the first row, and add -1 times the second row to the third row.

$$\begin{bmatrix} 0 & 5 & -7 \\ 1 & -1 & 2 \\ 0 & 2 & -8 \end{bmatrix}$$

Divide the third row by 2.

$$\begin{bmatrix} 0 & 5 & -7 \\ 1 & -1 & 2 \\ 0 & 1 & -4 \end{bmatrix}$$

Add -5 times the third row to the first row, and add the third row to the second row.

$$\begin{bmatrix} 0 & 0 & 13 \\ 1 & 0 & -2 \\ 0 & 1 & -4 \end{bmatrix}$$

Divide row one by 13; then add 2 times the resulting row one to row two. Next, add 4 times the resulting row one to row three.

$$\begin{bmatrix} 0 & 0 & 1 \\ 1 & 0 & 0 \\ 0 & 1 & 0 \end{bmatrix}$$

Interchange rows one and two, then rows two and three.

$$\begin{bmatrix} 1 & 0 & 0 \\ 0 & 1 & 0 \\ 0 & 0 & 1 \end{bmatrix}$$

This matrix is reduced and gives the system:

$$x = 0$$
$$y = 0$$
$$z = 0$$

Thus, the unique solution to the original system is $x = y = z = 0$, known as the trivial solution.

 Let $A = \begin{bmatrix} 2 & 1 & 4 \\ 3 & 0 & 1 \\ 2 & -1 & 1 \end{bmatrix}$ be the coefficient matrix of a

homogeneous system in x, y, and z. Solve this system to illustrate that a homogeneous system of three equations in the unknowns, x, y, z, has a unique solution.

A Perform the following row operations on matrix A.

Divide the first row by 2.

$$\begin{bmatrix} 1 & \frac{1}{2} & 2 \\ 3 & 0 & 1 \\ 2 & -1 & 1 \end{bmatrix}$$

Add -3 times the first row to the second row and -2 times the first row to the third row.

$$\begin{bmatrix} 1 & \frac{1}{2} & 2 \\ 0 & -\frac{3}{2} & -5 \\ 0 & -2 & -3 \end{bmatrix}$$

Add $-4/3$ times the second row to the third row.

$$\begin{bmatrix} 1 & \frac{1}{2} & 2 \\ 0 & -\frac{3}{2} & -5 \\ 0 & 0 & \frac{11}{3} \end{bmatrix}$$

Multiply the third row by $3/11$. Next, add -2 times the third row to the first row and 5 times the third row to the second row.

$$\begin{bmatrix} 1 & \frac{1}{2} & 0 \\ 0 & -\frac{3}{2} & 0 \\ 0 & 0 & 1 \end{bmatrix}$$

Multiply the second row by $-2/3$. Finally, add $-1/2$ times the second row to the first row.

$$\begin{bmatrix} 1 & 0 & 0 \\ 0 & 1 & 0 \\ 0 & 0 & 1 \end{bmatrix}$$

The corresponding system of equations is

$$x = 0$$
$$y = 0$$
$$z = 0$$

Thus, the system has a unique solution, and it is the trivial solution.

1.3 Matrices

A matrix is a rectangular array of numbers, called entries.

EXAMPLES

a) $\begin{bmatrix} 6 & 2 \\ 3 & 1 \\ 0 & 0 \end{bmatrix}$

b) $\begin{bmatrix} 3 \\ 1 \end{bmatrix}$

c) $[1 \ 7 \ 2 \ 1]$

A matrix with n rows and n columns is called a square matrix of order n.

EXAMPLE

$\begin{bmatrix} 2 & 10 & 1 \\ 6 & 2 & 9 \\ 3 & 3 & 7 \end{bmatrix}$ is a square matrix of order 3.

Two matrices are called equal if they have the same size and entries in corresponding positions are the same.

Entries starting at the top left and proceeding to the bottom right of a square matrix are said to be on the main diagonal of that matrix.

EXAMPLE

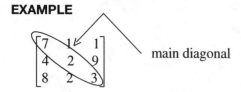
main diagonal

The sum $B + D$ is the matrix obtained when two matrices, B and D, are added together; they must both be of the same size. $B - D$ is obtained by subtracting the entries of D from the corresponding entries of B.

EXAMPLES

a) $\begin{bmatrix} 1 & 2 \\ 2 & 6 \end{bmatrix} + \begin{bmatrix} -4 & 7 \\ 1 & 1 \end{bmatrix} = \begin{bmatrix} -3 & 9 \\ 3 & 7 \end{bmatrix}$

b) $\begin{bmatrix} 1 & 2 \\ 2 & 6 \end{bmatrix} - \begin{bmatrix} -4 & 7 \\ 1 & 1 \end{bmatrix} = \begin{bmatrix} 5 & -5 \\ 1 & 5 \end{bmatrix}$

The product of a matrix A by a scalar k is obtained by multiplying each entry of A by k.

EXAMPLE

If $A = \begin{bmatrix} 4 & 7 \\ -1 & 2 \end{bmatrix}$ and $k = 3$, then $Ak = \begin{bmatrix} 12 & 21 \\ -3 & 6 \end{bmatrix}$.

When multiplying two matrices A and B, the matrices must be of the sizes $m \times n$ and $n \times p$ (the number of columns of A must equal the number of rows of B); to obtain the (ij) entry of AB, multiply the entries in row i of A by the corresponding entries in column j of B. Add up the resulting products; this sum is the (ij) entry of AB. If

$$AB = C, \text{ then } C_{ij} = \sum_{k=1}^{n} a_{ik}b_{kj} \,.$$

The size of C will be $m \times p$.

EXAMPLE

If $A = \begin{bmatrix} 2 & 3 \\ 4 & 5 \end{bmatrix}$ and $B = \begin{bmatrix} 3 & 3 \\ 7 & 2 \end{bmatrix}$, then

$$AB = \begin{bmatrix} 27 & 12 \\ 47 & 22 \end{bmatrix}.$$

A matrix which contains entries corresponding to the coefficients of a system of linear equations, but excludes the constants of that system, is called a coefficient matrix.

EXAMPLE

$$\begin{array}{rcl} x_1 + 6x_2 - 2x_3 &=& 4 \\ 3x_1 + x_3 &=& 7 \\ 5x_1 - 3x_2 + x_3 &=& 0 \end{array} \qquad \begin{bmatrix} 1 & 6 & -2 \\ 3 & 0 & 1 \\ 5 & -3 & 1 \end{bmatrix}$$

Problem Solving Examples:

Find $A + B$ where:

$$A = \begin{bmatrix} 1 & -2 & 4 \\ 2 & -1 & 3 \end{bmatrix}, \quad B = \begin{bmatrix} 0 & 2 & 4 \\ 1 & 3 & 1 \end{bmatrix}.$$

Using the definition of matrix addition, add the (ij) entry of A to the (ij) entry of B. Thus,

$$A + B = \begin{bmatrix} 1+0 & -2+2 & 4-4 \\ 2+1 & -1+3 & 3+1 \end{bmatrix} = \begin{bmatrix} 1 & 0 & 0 \\ 3 & 2 & 4 \end{bmatrix}$$

Let $A = \begin{bmatrix} 2 & 3 & 7 \\ 4 & m & \sqrt{3} \\ 1 & 5 & a \end{bmatrix}, \quad B = \begin{bmatrix} \alpha & \beta & \delta \\ \sqrt{5} & 3 & 1 \\ p & q & 4 \end{bmatrix}.$

Find $A + B$.

Using the definition of matrix addition, add the (ij) entry of A to the (ij) entry of B. Thus,

$$A + B = \begin{bmatrix} 2 & 3 & 7 \\ 4 & m & \sqrt{3} \\ 1 & 5 & a \end{bmatrix} + \begin{bmatrix} \alpha & \beta & \delta \\ \sqrt{5} & 3 & 1 \\ p & q & 4 \end{bmatrix}$$

$$= \begin{bmatrix} 2+\alpha & 3+\beta & 7+\delta \\ 4+\sqrt{5} & m+3 & \sqrt{3}+1 \\ 1+p & 5+q & a+4 \end{bmatrix}$$

If $A = \begin{bmatrix} 2 & 3 & 4 \\ 1 & 2 & 1 \end{bmatrix}$ and $B = \begin{bmatrix} 0 & 2 & 7 \\ 1 & -3 & 5 \end{bmatrix}$, find $A - B$.

$A - B$ is obtained by subtracting the entries of B from the corresponding entries of A.

$$A - B = \begin{bmatrix} 2 & 3 & 4 \\ 1 & 2 & 1 \end{bmatrix} - \begin{bmatrix} 0 & 2 & 7 \\ 1 & -3 & 5 \end{bmatrix}$$

$$= \begin{bmatrix} 2-0 & 3-2 & 4-7 \\ 1-1 & 2-(-3) & 1-5 \end{bmatrix}$$

$$= \begin{bmatrix} 2 & 1 & -3 \\ 0 & 5 & -4 \end{bmatrix}$$

Q If $A = \begin{bmatrix} 2 & -2 & 4 \\ -1 & 1 & 1 \end{bmatrix}$ and $B = \begin{bmatrix} 0 & 1 & -3 \\ 1 & 3 & 1 \end{bmatrix}$, find $2A + B$.

A $2A = 2 \begin{bmatrix} 2 & -2 & 4 \\ -1 & 1 & 1 \end{bmatrix}$

$$= \begin{bmatrix} 2 \times 2 & 2 \times (-2) & 2 \times 4 \\ 2 \times (-1) & 2 \times 1 & 2 \times 1 \end{bmatrix}$$

$$= \begin{bmatrix} 4 & -4 & 8 \\ -2 & 2 & 2 \end{bmatrix}$$

Then,

$$2A + B = \begin{bmatrix} 4 & -4 & 8 \\ -2 & 2 & 2 \end{bmatrix} + \begin{bmatrix} 0 & 1 & -3 \\ 1 & 3 & 1 \end{bmatrix}$$

$$= \begin{bmatrix} 4+0 & -4+1 & 8-3 \\ -2+1 & 2+3 & 2+1 \end{bmatrix}$$

$$2A + B = \begin{bmatrix} 4 & -3 & 5 \\ -1 & 5 & 3 \end{bmatrix}$$

Q If $A = \begin{bmatrix} 1 & 2 & 4 \\ 2 & 6 & 0 \end{bmatrix}$ and $B = \begin{bmatrix} 4 & 1 & 4 & 3 \\ 0 & -1 & 3 & 1 \\ 2 & 7 & 5 & 2 \end{bmatrix}$, find AB.

A Since A is a 2×3 matrix and B is a 3×4 matrix, the product AB is a 2×4 matrix.

$$AB = \begin{bmatrix} 1 & 2 & 4 \\ 2 & 6 & 0 \end{bmatrix} \begin{bmatrix} 4 & 1 & 4 & 3 \\ 0 & -1 & 3 & 1 \\ 2 & 7 & 5 & 2 \end{bmatrix}$$

$$= \begin{bmatrix} 1 \cdot 4 + 2 \cdot 0 + 4 \cdot 2 & 1 \cdot 1 + 2 \cdot (-1) + 4 \cdot 7 & 1 \cdot 4 + 2 \cdot 3 + 4 \cdot 5 & 1 \cdot 3 + 2 \cdot 1 + 4 \cdot 2 \\ 2 \cdot 2 + 6 \cdot 0 + 0 \cdot 2 & 2 \cdot 1 + 6 \cdot (-1) + 0 \cdot 7 & 2 \cdot 4 + 6 \cdot 3 + 0 \cdot 5 & 2 \cdot 3 + 6 \cdot 1 + 0 \cdot 2 \end{bmatrix}$$

$$= \begin{bmatrix} 4+0+8 & 1-2+28 & 4+6+20 & 3+2+8 \\ 8+0+0 & 2-6+0 & 8+18+0 & 6+6+0 \end{bmatrix}$$

$$AB = \begin{bmatrix} 12 & 27 & 30 & 13 \\ 8 & -4 & 26 & 12 \end{bmatrix}$$

1.4   Matrix Arithmetic

1.4.1   Rules of Matrix Arithmetic

a) $A + B = B + A$ (Commutative Law of Addition)

b) $A + (B + C) = (A + B) + C$ (Associative Law of Addition)

c) $A(BC) = (AB)C$ (Associative Law of Multiplication)

d) $A(B \pm C) = AB \pm AC$ (Distributive Law)

e) $a(B + C) = aB + aC$

f) $(a \pm b)C = aC \pm bC$

g) $(ab)C = a(bC)$

h) $a(BC) = (aB)C = B(aC)$

A matrix whose entries are all zero is called a zero matrix, **0**.

EXAMPLES

a) $\begin{bmatrix} 0 & 0 \\ 0 & 0 \end{bmatrix}$

b) $\begin{bmatrix} 0 \\ 0 \\ 0 \end{bmatrix}$

c) $\begin{bmatrix} 0 & 0 & 0 \\ 0 & 0 & 0 \\ 0 & 0 & 0 \end{bmatrix}$

THEOREM

If the size of the matrices are such that the indicated operations can be performed, the following rules of matrix arithmetic are valid:

a) $A + 0 = 0 + A = A$

b) $A - A = 0$

c) $0 - A = -A$

d) $A0 = 0$

An identity matrix (I) is a square matrix with ones on the main diagonal and zeros everywhere else.

EXAMPLES

a) $\begin{bmatrix} 1 & 0 \\ 0 & 1 \end{bmatrix}$

b) $\begin{bmatrix} 1 & 0 & 0 & 0 \\ 0 & 1 & 0 & 0 \\ 0 & 0 & 1 & 0 \\ 0 & 0 & 0 & 1 \end{bmatrix}$

If A is a square matrix and a matrix B exists such that $AB = BA = I$, then A is invertible and B is the inverse of A, (A^{-1}). An invertible matrix has one and only one inverse.

THEOREM

If A and B are invertible matrices of the same size, then:

a) AB is invertible

b) $(AB)^{-1} = (B^{-1})(A^{-1})$

The formula for inverting a 2×2 matrix is

$$\text{If } A = \begin{bmatrix} a & b \\ c & d \end{bmatrix}, \text{ then } A^{-1} = \frac{1}{ad - bc}\begin{bmatrix} d & -b \\ -c & a \end{bmatrix}.$$

EXAMPLE

$$\text{If } A = \begin{bmatrix} 1 & 2 \\ 3 & 4 \end{bmatrix}, \text{ then } A^{-1} = \begin{bmatrix} -2 & 1 \\ \frac{3}{2} & -\frac{1}{2} \end{bmatrix}.$$

THEOREM

If A is an invertible matrix, then:

a) A^{-1} is invertible; $(A^{-1})^{-1} = A$

b) kA is invertible (where k is a non-zero scalar); $(kA)^{-1} = \dfrac{1}{k} A^{-1}$

c) A^n is invertible; $(A^n)^{-1} = (A^{-1})^n$

If A is a square matrix and x and y are positive integers, then:

a) $A^x A^y = A^{x+y}$

b) $(A^x)^y = A^{xy}$

Problem Solving Examples:

Show that

a) $A + B = B + A$ where:

$$A = \begin{bmatrix} 3 & 1 & 1 \\ 2 & -1 & 1 \end{bmatrix} \text{ and } B = \begin{bmatrix} 4 & 2 & -1 \\ 0 & 0 & 2 \end{bmatrix}.$$

b) $(A + B) + C = A + (B + C)$ where

$$A = \begin{bmatrix} -2 & 6 \\ 2 & 1 \end{bmatrix}, B = \begin{bmatrix} 2 & 1 \\ 0 & 3 \end{bmatrix}, \text{ and } C = \begin{bmatrix} -1 & 0 \\ 7 & 2 \end{bmatrix}.$$

c)   If A and the zero matrix have the same size, then $A + 0 = A$ where:

$$A = \begin{bmatrix} 2 & 1 \\ 1 & 2 \end{bmatrix}.$$

d) $A + (-A) = 0$ where:

$$A = \begin{bmatrix} 2 & 1 \\ 1 & 2 \end{bmatrix}.$$

e) $(ab)A = a(bA)$ where $a = -5$, $b = 3$, and:

$$A = \begin{bmatrix} 6 & -1 & 0 \\ 1 & 2 & 1 \end{bmatrix}.$$

f)   Find B if $2A - 3B + C = 0$ where:

$$A = \begin{bmatrix} -1 & 3 \\ 0 & 0 \end{bmatrix} \text{ and } C = \begin{bmatrix} -2 & -1 \\ -1 & 1 \end{bmatrix}.$$

a) By the definition of matrix addition,

$$A + B = \begin{bmatrix} 3 & 1 & 1 \\ 2 & -1 & 1 \end{bmatrix} + \begin{bmatrix} 4 & 2 & -1 \\ 0 & 0 & 2 \end{bmatrix}$$

$$= \begin{bmatrix} 3+4 & 1+2 & 1+(-1) \\ 2+0 & -1+0 & 1+2 \end{bmatrix}$$

$$= \begin{bmatrix} 7 & 3 & 0 \\ 2 & -1 & 3 \end{bmatrix}$$

and

$$B + A = \begin{bmatrix} 4 & 2 & -1 \\ 0 & 0 & 2 \end{bmatrix} + \begin{bmatrix} 3 & 1 & 1 \\ 2 & -1 & 1 \end{bmatrix}$$

$$= \begin{bmatrix} 4+3 & 2+1 & -1+1 \\ 0+2 & 0+(-1) & 2+1 \end{bmatrix}$$

$$= \begin{bmatrix} 7 & 3 & 0 \\ 2 & -1 & 3 \end{bmatrix}$$

Thus, $A + B = B + A$.

b) $$A + B = \begin{bmatrix} -2 & 6 \\ 2 & 1 \end{bmatrix} + \begin{bmatrix} 2 & 1 \\ 0 & 3 \end{bmatrix}$$

$$= \begin{bmatrix} -2+2 & 6+1 \\ 2+0 & 1+3 \end{bmatrix}$$

$$= \begin{bmatrix} 0 & 7 \\ 2 & 4 \end{bmatrix}$$

and

$$(A + B) + C = \begin{bmatrix} 0 & 7 \\ 2 & 4 \end{bmatrix} + \begin{bmatrix} -1 & 0 \\ 7 & 2 \end{bmatrix}$$

$$= \begin{bmatrix} 0+(-1) & 7+0 \\ 2+7 & 4+2 \end{bmatrix}$$

$$= \begin{bmatrix} -1 & 7 \\ 9 & 6 \end{bmatrix}$$

$$B + C = \begin{bmatrix} 2 & 1 \\ 0 & 3 \end{bmatrix} + \begin{bmatrix} -1 & 0 \\ 7 & 2 \end{bmatrix}$$

$$= \begin{bmatrix} 2 + (-1) & 1 + 0 \\ 0 + 7 & 3 + 2 \end{bmatrix}$$

$$= \begin{bmatrix} 1 & 1 \\ 7 & 5 \end{bmatrix}$$

and

$$A + (B + C) = \begin{bmatrix} -2 & 6 \\ 2 & 1 \end{bmatrix} + \begin{bmatrix} 1 & 1 \\ 7 & 5 \end{bmatrix}$$

$$= \begin{bmatrix} -2 + 1 & 6 + 1 \\ 2 + 7 & 1 + 5 \end{bmatrix}$$

$$= \begin{bmatrix} -1 & 7 \\ 9 & 6 \end{bmatrix}$$

Thus, $(A + B) + C = A + (B + C)$.

c) $A = \begin{bmatrix} 2 & 1 \\ 1 & 2 \end{bmatrix}$ $\mathbf{0} = \begin{bmatrix} 0 & 0 \\ 0 & 0 \end{bmatrix}$.

Thus,

$$A + \mathbf{0} = \begin{bmatrix} 2 & 1 \\ 1 & 2 \end{bmatrix} + \begin{bmatrix} 0 & 0 \\ 0 & 0 \end{bmatrix}$$

$$= \begin{bmatrix} 2 + 0 & 1 + 0 \\ 1 + 0 & 2 + 0 \end{bmatrix}$$

$$= \begin{bmatrix} 2 & 1 \\ 1 & 2 \end{bmatrix}$$

Hence, $A + \mathbf{0} = A$.

d) $-A = -1 \times \begin{bmatrix} 2 & 1 \\ 1 & 2 \end{bmatrix}$

$$= \begin{bmatrix} -1 \times 2 & -1 \times 1 \\ -1 \times 1 & -1 \times 2 \end{bmatrix}$$

$$= \begin{bmatrix} -2 & -1 \\ -1 & -2 \end{bmatrix}$$

Thus,

$$A + (-A) = \begin{bmatrix} 2 & 1 \\ 1 & 2 \end{bmatrix} + \begin{bmatrix} -2 & -1 \\ -1 & -2 \end{bmatrix}$$

$$= \begin{bmatrix} 2 + (-2) & 1 + (-1) \\ 1 + (-1) & 2 + (-2) \end{bmatrix}$$

$$= \begin{bmatrix} 0 & 0 \\ 0 & 0 \end{bmatrix}$$

Therefore, $A + (-A) = \mathbf{0}$.

e) $\quad bA = 3 \begin{bmatrix} 6 & -1 & 0 \\ 1 & 2 & 1 \end{bmatrix} = \begin{bmatrix} 3 \times 6 & 3 \times (-1) & 3 \times 0 \\ 3 \times 1 & 3 \times 2 & 3 \times 1 \end{bmatrix}$

$$= \begin{bmatrix} 18 & -3 & 0 \\ 3 & 6 & 3 \end{bmatrix}$$

and

$$a(bA) = -5 \begin{bmatrix} 18 & -3 & 0 \\ 3 & 6 & 3 \end{bmatrix}$$

$$= \begin{bmatrix} -90 & 15 & 0 \\ -15 & -30 & -15 \end{bmatrix}$$

$$(ab)A = ((-5)(3)) \begin{bmatrix} 6 & -1 & 0 \\ 1 & 2 & 1 \end{bmatrix}$$

$$= -15 \begin{bmatrix} 6 & -1 & 0 \\ 1 & 2 & 1 \end{bmatrix}$$

$$= \begin{bmatrix} -90 & 15 & 0 \\ -15 & -30 & -15 \end{bmatrix}$$

Thus, $(ab)A = a(bA)$.

f) $2A - 3B + C = 2A + C - 3B = 0$ since matrix addition is commutative.

Now, add $3B$ to both sides of the equation,

$$2A + C - 3B = 0,$$

to obtain $2A + C - 3B + 3B = 0 + 3B$. (1)

Using the laws we exemplified in parts a) through d), (1) becomes $2A + C = 3B$. Now,

$$\frac{1}{3}(2A + C) = \frac{1}{3}(3B)$$

which implies

$$B = \frac{1}{3}(2A + C).$$

$$2A + C = \begin{bmatrix} 2(-1) & 2(3) \\ 2(0) & 2(0) \end{bmatrix} + \begin{bmatrix} -2 & -1 \\ -1 & 1 \end{bmatrix} = \begin{bmatrix} -4 & 5 \\ -1 & 1 \end{bmatrix}$$

Thus,

$$B = \frac{1}{3}(2A + C) = \frac{1}{3}\begin{bmatrix} -4 & 5 \\ -1 & 1 \end{bmatrix} = \begin{bmatrix} -\frac{4}{3} & \frac{5}{3} \\ -\frac{1}{3} & \frac{1}{3} \end{bmatrix}.$$

Q Let $A = \begin{bmatrix} 1 & 1 \\ 3 & 7 \end{bmatrix}$ and $B = \begin{bmatrix} 2 & 5 \\ 4 & 0 \end{bmatrix}$. Show $AB \neq BA$.

A
$$AB = \begin{bmatrix} 1 & 1 \\ 3 & 7 \end{bmatrix}\begin{bmatrix} 2 & 5 \\ 4 & 0 \end{bmatrix} = \begin{bmatrix} 1 \cdot 2 + 1 \cdot 4 & 1 \cdot 5 + 1 \cdot 0 \\ 3 \cdot 2 + 7 \cdot 4 & 3 \cdot 5 + 7 \cdot 0 \end{bmatrix}$$

$$= \begin{bmatrix} 2 + 4 & 5 + 0 \\ 6 + 28 & 15 + 0 \end{bmatrix}$$

$$= \begin{bmatrix} 6 & 5 \\ 34 & 15 \end{bmatrix}$$

$$BA = \begin{bmatrix} 2 & 5 \\ 4 & 0 \end{bmatrix}\begin{bmatrix} 1 & 1 \\ 3 & 7 \end{bmatrix} = \begin{bmatrix} 2 \cdot 1 + 5 \cdot 3 & 2 \cdot 1 + 5 \cdot 7 \\ 4 \cdot 1 + 0 \cdot 3 & 4 \cdot 1 + 0 \cdot 7 \end{bmatrix}$$

$$= \begin{bmatrix} 2 + 15 & 2 + 35 \\ 4 + 0 & 4 + 0 \end{bmatrix}$$

$$= \begin{bmatrix} 17 & 37 \\ 4 & 4 \end{bmatrix}$$

Therefore, $AB \neq BA$.

Find (a) A^2, (b) A^3, (c) A^4, when $A = \begin{bmatrix} 1 & 2 \\ -1 & 1 \end{bmatrix}$.

a) $A^2 = AA = \begin{bmatrix} 1 & 2 \\ -1 & 1 \end{bmatrix}\begin{bmatrix} 1 & 2 \\ -1 & 1 \end{bmatrix} = \begin{bmatrix} 1-2 & 2+2 \\ -1-1 & -2+1 \end{bmatrix} = \begin{bmatrix} -1 & 4 \\ -2 & -1 \end{bmatrix}$.

(b) $A^3 = AAA = A^2A = \begin{bmatrix} -1 & 4 \\ -2 & -1 \end{bmatrix}\begin{bmatrix} 1 & 2 \\ -1 & 1 \end{bmatrix} = \begin{bmatrix} -1-4 & -2+4 \\ -2+1 & -4-1 \end{bmatrix} = \begin{bmatrix} -5 & 2 \\ -1 & -5 \end{bmatrix}$.

The usual laws for exponents are $A^m A^n = A^{m+n}$ and $(A^m)^n = A^{mn}$. Thus, $A^4 = A^3A$ or, $A^4 = (A^2)^2 = A^2 A^2$.

(c) $A^4 = A^3A = \begin{bmatrix} -5 & 2 \\ -1 & -5 \end{bmatrix}\begin{bmatrix} 1 & 2 \\ -1 & 1 \end{bmatrix} = \begin{bmatrix} -5-2 & -10+2 \\ -1+5 & -2-5 \end{bmatrix} = \begin{bmatrix} -7 & -8 \\ 4 & -7 \end{bmatrix}$.

Observe that

$$A^4 = A^2 A^2 = \begin{bmatrix} -1 & 4 \\ -2 & -1 \end{bmatrix}\begin{bmatrix} -1 & 4 \\ -2 & -1 \end{bmatrix} = \begin{bmatrix} 1-8 & -4-4 \\ 2+2 & -8+1 \end{bmatrix}$$

$$= \begin{bmatrix} -7 & -8 \\ 4 & -7 \end{bmatrix}$$

1.5 Gaussian Elimination

A matrix is in reduced row-echelon form if it has the following properties:

a) Either the first non-zero entry in a row is one, or the row consists entirely of zeros.

b) All rows consisting entirely of zeros are grouped together at the bottom of the matrix.

c) If two successive rows do not consist entirely of zeros, then the leading one in the lower row occurs farther to the right than the leading one in the higher row.

d) Every column with a leading one has zeros in every other entry.

If a matrix has properties a, b, and c, it is in row-echelon form.

EXAMPLES

a) The matrix $\begin{bmatrix} 1 & 0 & 0 & 4 \\ 0 & 1 & 1 & 2 \\ 0 & 0 & 0 & 1 \end{bmatrix}$ is in row-echelon form.

b) The matrix $\begin{bmatrix} 1 & 0 & 0 & 3 \\ 0 & 1 & 0 & 0 \\ 0 & 0 & 1 & 2 \end{bmatrix}$ is in reduced row-echelon form.

The variables corresponding to the leading ones in a reduced row-echelon matrix are called leading variables. Gaussian elimination is a process using elementary row operations by which any matrix can be brought into reduced row-echelon form. Once this is done the system of linear equations corresponding to that matrix is easily solvable. The process is as follows:

a) Find the leftmost column of the matrix not consisting entirely of zeros.

b) If necessary, switch the top row with another row so that a non-zero entry appears at the top of the column found in (a).

c) If necessary, multiply the top row by the multiplicative inverse of the entry in that row which is also the top of the column found in (a). This is done so that this column has a leading one.

d) Add appropriate multiples of the first row to the rows below so that all entries below the leading one of the column found in (a) become zeros.

e) Cover the first row and, using the remaining submatrix, begin again at step (a). Continue (a) through (e) until the matrix is in row-echelon form.

f) Starting with the last row which does not consist entirely of zeros, add appropriate multiples of each row to the rows above so that each column containing a one has zeros everywhere else. The matrix will now be in reduced row-echelon form.

EXAMPLE

The system

$$x + y = 1$$
$$6x - 2z = -8,$$
$$3y - z = -3,$$

corresponds to the augmented matrix

$$\left[\begin{array}{ccc|c} 1 & 1 & 0 & 1 \\ 6 & 0 & -2 & -8 \\ 0 & 3 & -1 & -3 \end{array}\right].$$

This matrix may be reduced to the matrix $\left[\begin{array}{ccc|c} 1 & 0 & 0 & \frac{1}{3} \\ 0 & 1 & 0 & \frac{2}{3} \\ 0 & 0 & 1 & 5 \end{array}\right]$ which

corresponds to the system

$$x = \frac{1}{3}$$
$$y = \frac{2}{3}$$
$$z = 5,$$

and the system is solved.

Problem Solving Examples:

 Given

$$A = \left[\begin{array}{ccc|c} 1 & -2 & 3 & -1 \\ 2 & -1 & 2 & 2 \\ 3 & 1 & 2 & 3 \end{array}\right],$$

a) Reduce A to echelon form.

b) Reduce A to row-reduced echelon form.

a) To reduce A to echelon form, apply the following row operations:

Add –2 times the first row to the second row and –3 times the first row to the third row.

$$\left[\begin{array}{ccc|c} 1 & -2 & 3 & -1 \\ 0 & 3 & -4 & 4 \\ 0 & 7 & -7 & 6 \end{array}\right]$$

Multiply the second row by 7 and the third row by 3.

$$\left[\begin{array}{ccc|c} 1 & -2 & 3 & -1 \\ 0 & 21 & -28 & 28 \\ 0 & 21 & -21 & 18 \end{array}\right]$$

Then, add –1 times the second row to the third row, to obtain

$$\left[\begin{array}{ccc|c} 1 & -2 & 3 & -1 \\ 0 & 3 & -4 & 4 \\ 0 & 0 & 7 & -10 \end{array}\right]$$

Divide the second row by 3 and the third row by 7 to obtain the echelon form.

$$\left[\begin{array}{ccc|c} 1 & -2 & 3 & -1 \\ 0 & 1 & -\frac{4}{3} & \frac{4}{3} \\ 0 & 0 & 1 & -\frac{10}{7} \end{array}\right]$$

b) To obtain the reduced echelon form, add 2 times the second row to the first row.

$$\left[\begin{array}{ccc|c} 1 & 0 & \frac{1}{3} & \frac{5}{3} \\ 0 & 1 & -\frac{4}{3} & \frac{4}{3} \\ 0 & 0 & 1 & -\frac{10}{7} \end{array}\right]$$

Add $-\frac{1}{3}$ times the third row to the first row and $\frac{4}{3}$ times the third row to the second row, resulting in the row reduced echelon form.

$$\begin{bmatrix} 1 & 0 & 0 & | & \frac{15}{7} \\ 0 & 1 & 0 & | & -\frac{4}{7} \\ 0 & 0 & 1 & | & -\frac{10}{7} \end{bmatrix}$$

 Solve the following systems of equations:

(1) $\quad 4x_1 - 3x_2 + x_3 = -1$
$\quad\quad x_1 + 5x_2 - 2x_3 = 2$
$\quad\quad x_1 + 2x_2 \quad\quad = 0$

(2) $\quad 2x_1 + 3x_2 + x_3 - 4x_4 = 0$
$\quad\quad x_1 - 5x_2 - 3x_3 + 2x_4 = 0$
$\quad\quad 5x_1 + 2x_2 \quad\quad - x_4 = 0$
$\quad\quad 2x_1 - 9x_2 - 5x_3 + 9x_4 = 0$

 (1) The augmented matrix for system (1) is:

$$\begin{bmatrix} 4 & -3 & 1 & | & -1 \\ 1 & 5 & -2 & | & 2 \\ 1 & 2 & 0 & | & 0 \end{bmatrix}. \qquad (1)$$

Reduce this matrix to reduced row-echelon form.

Interchange the first and the third rows.

$$\begin{bmatrix} 1 & 2 & 0 & | & 0 \\ 1 & 5 & -2 & | & 2 \\ 4 & -3 & 1 & | & -1 \end{bmatrix}. \qquad (2)$$

Add -1 times the first row to the second row and -4 times the first row to the third row.

$$\begin{bmatrix} 1 & 2 & 0 & | & 0 \\ 0 & 3 & -2 & | & 2 \\ 0 & -11 & 1 & | & -1 \end{bmatrix}. \qquad (3)$$

Divide the second row by 3 and the third row by –11.

$$\begin{bmatrix} 1 & 2 & 0 & | & 0 \\ 0 & 1 & -\frac{2}{3} & | & \frac{2}{3} \\ 0 & 1 & -\frac{1}{11} & | & \frac{1}{11} \end{bmatrix} \tag{4}$$

Add –1 times the second row to the third row to obtain the row-echelon form.

$$\begin{bmatrix} 1 & 2 & 0 & | & 0 \\ 0 & 1 & -\frac{2}{3} & | & \frac{2}{3} \\ 0 & 0 & \frac{19}{33} & | & -\frac{19}{33} \end{bmatrix}. \tag{5}$$

It can be seen easily that system (1) has a unique solution because system (1) has three unknowns and the row-echelon matrix (5) has three non-zero rows. Matrix (5) is the augmented matrix of the system.

$$\begin{aligned} x_1 + 2x_2 \quad\quad &= 0 \\ x_2 - \tfrac{2}{3} x_3 &= \tfrac{2}{3} \\ \tfrac{19}{33} x_3 &= -\tfrac{19}{33}. \end{aligned}$$

Solving this system yields $x_1 = 0$, $x_2 = 0$, and $x_3 = -1$.

(2)   The matrix of coefficients for system (2) is:

$$\begin{bmatrix} 2 & 3 & 1 & -4 \\ 1 & -5 & -3 & 2 \\ 5 & 2 & 0 & -1 \\ 2 & -9 & -5 & 9 \end{bmatrix}.$$

Interchange the first and the second rows.

$$\begin{bmatrix} 1 & -5 & -3 & 2 \\ 2 & 3 & 1 & -4 \\ 5 & 2 & 0 & -1 \\ 2 & -9 & -5 & 9 \end{bmatrix}$$

Add –2 times the first row to the second row and to the fourth row. Add –5 times the first row to the third row.

$$\begin{bmatrix} 1 & -5 & -3 & 2 \\ 0 & 13 & 7 & -8 \\ 0 & 27 & 15 & -11 \\ 0 & 1 & 1 & 5 \end{bmatrix}$$

Add 5 times the fourth row to the first row and –13 times the fourth row to the second row. Add –27 times the fourth row to the third row.

$$\begin{bmatrix} 1 & 0 & 2 & 27 \\ 0 & 0 & -6 & -73 \\ 0 & 0 & -12 & -146 \\ 0 & 1 & 1 & 5 \end{bmatrix}$$

Add –2 times the second row to the third row.

$$\begin{bmatrix} 1 & 0 & 2 & 27 \\ 0 & 0 & -6 & -73 \\ 0 & 0 & 0 & 0 \\ 0 & 1 & 1 & 5 \end{bmatrix}$$

Interchange rows two and four, then rows three and four.

$$\begin{bmatrix} 1 & 0 & 2 & 27 \\ 0 & 1 & 1 & 5 \\ 0 & 0 & -6 & -73 \\ 0 & 0 & 0 & 0 \end{bmatrix}$$

Divide the third row by –6. Then add –1 times the third row to the second row and –2 times the third row to the first row.

$$\begin{bmatrix} 1 & 0 & 0 & \frac{8}{3} \\ 0 & 1 & 0 & -\frac{43}{6} \\ 0 & 0 & 1 & \frac{73}{6} \\ 0 & 0 & 0 & 0 \end{bmatrix}$$

This is the matrix of coefficients for the system:

$$x_1 \quad + \; \tfrac{8}{3}\,x_4 = 0$$
$$x_2 \; - \tfrac{43}{6}\,x_4 = 0$$
$$x_3 + \tfrac{73}{6}\,x_4 = 0$$

The above homogeneous system has a non-trivial solution since the system involves more unknowns than equations. Solving for the leading variables yields:

$$x_1 = -\tfrac{8}{3}\,x_4$$
$$x_2 = \tfrac{43}{6}\,x_4$$
$$x_3 = -\tfrac{73}{6}\,x_4$$

The solution set is, therefore, given by $x_1 = -\tfrac{8}{3}\,t$, $x_2 = \tfrac{43}{6}\,t$, $x_3 = -\tfrac{73}{6}\,t$, and $x_4 = t$.

Q Suppose that the augmented matrix for a system of linear equations has been reduced by row operations to the given reduced row-echelon form. Solve the system.

(a) $\begin{bmatrix} 1 & 0 & 0 & | & 5 \\ 0 & 1 & 0 & | & -2 \\ 0 & 0 & 1 & | & 4 \end{bmatrix}$

(b) $\begin{bmatrix} 1 & 0 & 0 & 4 & | & -1 \\ 0 & 1 & 0 & 2 & | & 6 \\ 0 & 0 & 1 & 3 & | & 2 \end{bmatrix}$

(c) $\begin{bmatrix} 1 & 6 & 0 & 0 & 4 & | & -2 \\ 0 & 0 & 1 & 0 & 3 & | & 1 \\ 0 & 0 & 0 & 1 & 5 & | & 2 \\ 0 & 0 & 0 & 0 & 0 & | & 0 \end{bmatrix}$

(d) $\begin{bmatrix} 1 & 0 & 0 & | & 0 \\ 0 & 1 & 0 & | & 0 \\ 0 & 0 & 0 & | & 1 \end{bmatrix}$

A (a) The corresponding system of equations is:

$$x_1 = 5$$
$$x_2 = -2$$
$$x_3 = 4$$

Therefore, the solution to the system is:

$$x_1 = 5, x_2 = -2, \text{ and } x_3 = 4.$$

(b) The corresponding system of equations is:

$$x_1 \qquad + 4x_4 = -1$$
$$x_2 + 2x_4 = 6$$
$$x_3 + 3x_4 = 2$$

The above system can be written as:

$$x_1 = -1 - 4x_4$$
$$x_2 = 6 - 2x_4 \qquad\qquad (1)$$
$$x_3 = 2 - 3x_4$$

Assign x_4 any value, and then compute x_1, x_2, and x_3 from (1). Thus, we have many solutions, one for each value of x_4.

(c) The corresponding system of equations is:

$$x_1 + 6x_2 + 4x_5 = -2$$
$$x_3 + 3x_5 = 1$$
$$x_4 + 5x_5 = 2$$

The above equations can be written as:

$$x_1 = -2 - 4x_5 - 6x_2$$
$$x_3 = 1 - 3x_5$$
$$x_4 = 2 - 5x_5$$

Since x_5 can be assigned an arbitrary value, t, and x_2 can be assigned an arbitrary value, s, there are infinitely many solutions. The solution set is given by the formula:

$$x_1 = -2 - 4t - 6s, \ x_2 = s, \ x_3 = 1 - 3t,$$
$$x_4 = 2 - 5t, \ x_5 = 5$$

(d) This system has no solution since the reduced row-echelon form has a row in which the first non-zero entry is in the last column.

1.6 Elementary Matrices

An $n \times n$ matrix, which can be derived from the $n \times n$ identity matrix by performing a single elementary row operation, is called an elementary matrix (E).

THEOREM

If E is an $m \times m$ elementary matrix and B is an $m \times n$ matrix, then the product of EB is equivalent to performing the row operation of E on B.

THEOREM

Every elementary matrix is invertible, and its inverse is also the elementary matrix.

Row-equivalent matrices are matrices which can be derived from each other by a finite sequence of row operations.

THEOREM

If A is an $n \times n$ matrix, then the following statements are equivalent (either all true or all false):

a) A is invertible.

b) The system of linear equations represented by $AX = 0$ has only the trivial solution.

c) A is row-equivalent to the $n \times n$ identity matrix.

To invert an $n \times n$ matrix, A, where $n > 2$, you must perform the elementary row operations on the $n \times n$ identity matrix which would reduce A to I. That derivation of the $n \times n$ identity matrix will be A^{-1}.

THEOREM

If A is an invertible $n \times n$ matrix, then for the system of equations, $AX = B$, where B is any $n \times 1$ matrix, there is only one solution; namely $X = A^{-1}B$.

THEOREM

If A is an $n \times n$ matrix, then the following statements are equivalent:

a) A is invertible.

b) $AX = 0$ has only the trivial solution.

c) *A* is row-equivalent to the $n \times n$ identity matrix.

d) *AX* = *B* is consistent for every $n \times n$ matrix *B*.

Problem Solving Examples:

Find the inverse of *A* where:

$$A = \begin{bmatrix} 1 & 1 & 1 \\ 0 & 2 & 3 \\ 5 & 5 & 1 \end{bmatrix}.$$

We first form the matrix $[A : I]$.

$$\begin{bmatrix} 1 & 1 & 1 & | & 1 & 0 & 0 \\ 0 & 2 & 3 & | & 0 & 1 & 0 \\ 5 & 5 & 1 & | & 0 & 0 & 1 \end{bmatrix}$$

We arrange our computations as follows:

$$\begin{bmatrix} 1 & 1 & 1 & | & 1 & 0 & 0 \\ 0 & 2 & 3 & | & 0 & 1 & 0 \\ 5 & 5 & 1 & | & 0 & 0 & 1 \end{bmatrix}$$

Subtract 5 times the first row from the third row to obtain:

$$\begin{bmatrix} 1 & 1 & 1 & | & 1 & 0 & 0 \\ 0 & 2 & 3 & | & 0 & 1 & 0 \\ 0 & 0 & -4 & | & -5 & 0 & 1 \end{bmatrix}$$

Divide the second row by 2 to obtain:

$$\begin{bmatrix} 1 & 1 & 1 & | & 1 & 0 & 0 \\ 0 & 1 & \frac{3}{2} & | & 0 & \frac{1}{2} & 0 \\ 0 & 0 & -4 & | & -5 & 0 & 1 \end{bmatrix}$$

Subtract the second row from the first row to obtain:

$$\begin{bmatrix} 1 & 0 & -\frac{1}{2} & 1 & -\frac{1}{2} & 0 \\ 0 & 1 & \frac{3}{2} & 0 & \frac{1}{2} & 0 \\ 0 & 0 & -4 & -5 & 0 & 1 \end{bmatrix}.$$

Divide the third row by −4 to obtain:

$$\begin{bmatrix} 1 & 0 & -\frac{1}{2} & 1 & -\frac{1}{2} & 0 \\ 0 & 0 & \frac{3}{2} & 0 & \frac{1}{2} & 0 \\ 0 & 0 & 1 & \frac{5}{4} & 0 & -\frac{1}{4} \end{bmatrix}.$$

Add $-\frac{3}{2}$ times the third row to the second row to obtain:

$$\begin{bmatrix} 1 & 0 & -\frac{1}{2} & 1 & -\frac{1}{2} & 0 \\ 0 & 1 & 0 & -\frac{15}{8} & \frac{1}{2} & \frac{3}{8} \\ 0 & 0 & 1 & \frac{5}{4} & 0 & -\frac{1}{4} \end{bmatrix}.$$

Add $\frac{1}{2}$ times the third row to the first row to obtain:

$$\begin{bmatrix} 1 & 0 & 0 & \frac{13}{8} & -\frac{1}{2} & -\frac{1}{8} \\ 0 & 1 & 0 & -\frac{15}{8} & \frac{1}{2} & \frac{3}{8} \\ 0 & 0 & 1 & \frac{5}{4} & 0 & -\frac{1}{4} \end{bmatrix}.$$

Hence,

$$A^{-1} = \begin{bmatrix} \frac{13}{8} & -\frac{1}{2} & -\frac{1}{8} \\ -\frac{15}{8} & \frac{1}{2} & \frac{3}{8} \\ \frac{5}{4} & 0 & -\frac{1}{4} \end{bmatrix}.$$

 Solve the following system:

$$\begin{aligned} x_1 - 2x_2 - 3x_3 &= 3 \\ 2x_1 - x_2 - 4x_3 &= 7 \\ 3x_1 - 3x_2 - 5x_3 &= 8 \end{aligned} \qquad (1)$$

The matrix of coefficients for system (1) is:

$$A = \begin{bmatrix} 1 & -2 & -3 \\ 2 & -1 & -4 \\ 3 & -3 & -5 \end{bmatrix}.$$

System (1) may be written in matrix form as:

$$\begin{bmatrix} 1 & -2 & -3 \\ 2 & -1 & -4 \\ 3 & -3 & -5 \end{bmatrix} \begin{bmatrix} x_1 \\ x_2 \\ x_3 \end{bmatrix} = \begin{bmatrix} 3 \\ 7 \\ 8 \end{bmatrix}. \tag{2}$$

Let:

$$X = \begin{bmatrix} x_1 \\ x_2 \\ x_3 \end{bmatrix} \text{ and } B = \begin{bmatrix} 3 \\ 7 \\ 8 \end{bmatrix}.$$

Then equation (2) is written:

$$AX = B. \tag{3}$$

The solution is found by multiplying both sides of equation (3) by A^{-1}.

Thus,

$$X = A^{-1}B. \tag{4}$$

To find A^{-1}, first form the matrix $[A : I]$, and reduce this matrix by applying row operations to the form $[I : B]$. Then, $B = A^{-1}$. Now,

$$[A:I] = \begin{bmatrix} 1 & -2 & -3 & | & 1 & 0 & 0 \\ 2 & -1 & -4 & | & 0 & 1 & 0 \\ 3 & -3 & -5 & | & 0 & 0 & 1 \end{bmatrix}.$$

Add −2 times the first row to the second row and −3 times the first row to the third row.

$$\left[\begin{array}{ccc|ccc} 1 & -2 & -3 & 1 & 0 & 0 \\ 0 & 3 & 2 & -2 & 1 & 0 \\ 0 & 3 & 4 & -3 & 0 & 1 \end{array}\right]$$

Now add −1 times the second row to the third row.

$$\left[\begin{array}{ccc|ccc} 1 & -2 & -3 & 1 & 0 & 0 \\ 0 & 3 & 2 & -2 & 1 & 0 \\ 0 & 0 & 2 & -1 & -1 & 0 \end{array}\right]$$

Divide the third row by 2.

$$\left[\begin{array}{ccc|ccc} 1 & -2 & -3 & 1 & 0 & 0 \\ 0 & 3 & 2 & -2 & 1 & 0 \\ 0 & 0 & 1 & -\frac{1}{2} & -\frac{1}{2} & \frac{1}{2} \end{array}\right]$$

Add −2 times the third row to the second row and 3 times the third row to the first row.

$$\left[\begin{array}{ccc|ccc} 1 & -2 & 0 & -\frac{1}{2} & -\frac{3}{2} & \frac{3}{2} \\ 0 & 3 & 0 & -1 & 2 & -1 \\ 0 & 0 & 1 & -\frac{1}{2} & -\frac{1}{2} & \frac{1}{2} \end{array}\right]$$

Divide the second row by 3; then add 2 times the resulting second row to the first row.

$$\left[\begin{array}{ccc|ccc} 1 & 0 & 0 & -\frac{7}{6} & -\frac{1}{6} & \frac{5}{6} \\ 0 & 1 & 0 & -\frac{1}{3} & \frac{2}{3} & -\frac{1}{3} \\ 0 & 0 & 1 & -\frac{1}{2} & -\frac{1}{2} & \frac{1}{2} \end{array}\right]$$

Thus,

$$A^{-1} = \left[\begin{array}{ccc} -\frac{7}{6} & -\frac{1}{6} & \frac{5}{6} \\ -\frac{1}{3} & \frac{2}{3} & -\frac{1}{3} \\ -\frac{1}{2} & -\frac{1}{2} & \frac{1}{2} \end{array}\right].$$

Then equation (4) becomes:

$$\begin{bmatrix} x_1 \\ x_2 \\ x_3 \end{bmatrix} = \begin{bmatrix} -\frac{7}{6} & -\frac{1}{6} & \frac{5}{6} \\ -\frac{1}{3} & \frac{2}{3} & -\frac{1}{3} \\ -\frac{1}{2} & -\frac{1}{2} & \frac{1}{2} \end{bmatrix} \begin{bmatrix} 3 \\ 7 \\ 8 \end{bmatrix}.$$

Multiplying, we have:

$$\begin{bmatrix} x_1 \\ x_2 \\ x_3 \end{bmatrix} = \begin{bmatrix} -\frac{21}{6} & -\frac{7}{6} & \frac{40}{6} \\ -1 & \frac{14}{3} & -\frac{8}{3} \\ -\frac{3}{2} & -\frac{7}{2} & 4 \end{bmatrix} = \begin{bmatrix} 2 \\ 1 \\ -1 \end{bmatrix}.$$

Thus,

$$x_1 = 2, \ x_2 = 1, \text{ and } x_3 = -1.$$

Quiz: Linear Matrices

1. The inverse of the matrix $M = \begin{bmatrix} 2 & 1 & 3 \\ 0 & -1 & 2 \\ 4 & 3 & 1 \end{bmatrix}$ is the matrix

    $M^{-1} = \dfrac{1}{6} \begin{bmatrix} -7 & 8 & a \\ 8 & -10 & -4 \\ 4 & b & -2 \end{bmatrix}$ where

    (A) $a = 5; b = -2.$ (D) $a = 2; b = -3.$

    (B) $a = 3; b = 2.$ (E) $a = 2; b = 3.$

    (C) $a = 1; b = -3.$

2. A system of linear equations $a_{ij}x_j, i = 1, \ldots, n, j = 1, \ldots, m$ has a non-trivial solution if

    (A) $n = m.$ (B) $n < m.$

(C) $m < n.$ (D) $mn = 1.$

(E) $mn = 0.$

3. The system of equations

$$2x + 3y - z = 3$$
$$x - 2y + 2z = 5$$
$$4x + 6y - 2z = 1$$

(A) has an infinite number of solutions.

(B) has no solutions.

(C) has a unique solution.

(D) has a non-trivial solution.

(E) has only the trivial solution.

4. For matrices

$$A = \begin{bmatrix} 1 & 1 \\ 0 & -1 \end{bmatrix}, \quad B = \begin{bmatrix} 0 & 1 \\ 1 & 1 \end{bmatrix}, \quad C = \begin{bmatrix} -1 & 0 \\ 1 & 1 \end{bmatrix},$$
$$\text{and } D = \begin{bmatrix} 3 & 3 \\ 0 & -2 \end{bmatrix},$$

the matrix D is a linear combination $(aA + bB + cC)$ of A, B, C for a, b, c given by

(A) 1, 1, –1.   (D) 1, –2, 1.

(B) 2, 1, –1.   (E) –1, 1, –2.

(C) 2, 2, –2.

5. If a homogeneous system of three equations has the solution (4, –2, 7), it will also have the solution

(A) (9, 3, 12).   (B) $\left(\dfrac{1}{4}, -\dfrac{1}{2}, \dfrac{1}{7} \right).$

(C) $(-20, 10, -35)$.   (D) $(-8, -14, 28)$.

(E) $(-2, -8, 1)$.

6. If $A = \begin{bmatrix} 1 & 2 \\ -3 & 0 \end{bmatrix}$ and $B = \begin{bmatrix} 3 & -1 \\ 4 & 2 \end{bmatrix}$, then $3A + 7B - 2AB$ is equal to

(A) $\begin{bmatrix} -1 & 5 \\ -1 & 8 \end{bmatrix}$.   (D) $\begin{bmatrix} -18 & 13 \\ -37 & -14 \end{bmatrix}$.

(B) $\begin{bmatrix} 1 & -5 \\ 1 & -8 \end{bmatrix}$.   (E) $\begin{bmatrix} 1 & 5 \\ 1 & 8 \end{bmatrix}$.

(C) $\begin{bmatrix} 24 & -1 \\ 19 & 14 \end{bmatrix}$.

7. Which of the matrices given below is the reduced row-echelon form of the following matrix?

$$\begin{bmatrix} 1 & 2 & -1 & 2 \\ 2 & 0 & 3 & 1 \\ 3 & 2 & 1 & 3 \end{bmatrix}$$

(A) $\begin{bmatrix} 0 & 0 & 1 & \frac{1}{2} \\ 0 & 1 & 0 & \frac{3}{4} \\ 1 & 0 & 0 & 0 \end{bmatrix}$ (D) $\begin{bmatrix} 1 & 0 & 0 & \frac{3}{4} \\ 0 & 1 & 0 & \frac{1}{2} \\ 0 & 0 & 1 & 0 \end{bmatrix}$

(B) $\begin{bmatrix} 1 & 0 & 0 & \frac{1}{2} \\ 1 & 0 & 0 & \frac{3}{4} \\ 1 & 0 & 0 & 0 \end{bmatrix}$ (E) $\begin{bmatrix} 1 & 0 & 0 & \frac{1}{2} \\ 0 & 1 & 0 & \frac{3}{4} \\ 0 & 0 & 1 & 0 \end{bmatrix}$

(C) $\begin{bmatrix} 1 & 0 & 0 & 0 \\ 0 & 1 & 0 & \frac{3}{4} \\ 0 & 0 & 1 & \frac{1}{2} \end{bmatrix}$

8. Given the matrix $A = \begin{bmatrix} 1 & 2 & 3 \\ 2 & 3 & 4 \\ 1 & 1 & 5 \end{bmatrix}$ and $B = A^{-1}$, find the entry in row 3 and column 2 of B.

(A) $\dfrac{1}{4}$.

(D) $-\dfrac{1}{2}$.

(B) $\dfrac{1}{2}$.

(E) $-\dfrac{1}{4}$.

(C) 1.

9. If the trace of a square matrix is defined to be the sum of the elements on the main diagonal, the trace of M^5, where $M = \begin{bmatrix} 6 & 10 \\ -2 & -3 \end{bmatrix}$, is which of the following?

(A) 27.

(D) $6^5 + (-3)^5$.

(B) 3^5.

(E) 33.

(C) 5^3.

10. Find the solution to the given system of equations by finding the inverse of the coefficient matrix.

$$2x + 3y - z = -5$$
$$-x + \tfrac{1}{2}y + 3z = 10$$
$$x + 5y - 7z = -22$$

(A) $(3, -2, 0)$.

(D) $(42, -56, 199)$.

(B) $(36, 6, -4)$.

(E) $(-1, 0, 3)$.

(C) $(2, 0, 4)$.

ANSWER KEY

1. (A)	6. (A)
2. (B)	7. (B)
3. (B)	8. (E)
4. (B)	9. (E)
5. (C)	10. (E)

CHAPTER 2

Determinants

2.1 Determinant Function

A permutation of a set of integers is some arrangement of those integers without any repetitions or omissions.

EXAMPLE

The permutations of $\{2, 3, 4\}$ are:

$(2, 3, 4), (2, 4, 3), (3, 2, 4), (3, 4, 2), (4, 2, 3), (4, 3, 2)$.

An inversion in a permutation occurs when a larger integer appears before a smaller one.

EXAMPLE

$(5, 2, 3, 7)$ There are two inversions.

An even permutation has an even number of inversions; an odd permutation has an odd number of inversions.

An elementary product from an $n \times n$ matrix A is a product of n entries from A, with no two entries from the same row or column.

EXAMPLE

$\begin{bmatrix} 1 & 2 \\ 3 & 4 \end{bmatrix}$ The elementary products are 4 and 6.

A signed elementary product from the matrix A is an elementary product of A multiplied by -1 or $+1$. We use the "+" sign if the permutation of the set is even and the "–" if odd.

EXAMPLE

$\begin{bmatrix} 1 & 4 \\ 3 & 2 \end{bmatrix}$ The signed elementary products are 2 and -12.

The determinant of a square matrix A ($\det(A)$) is the sum of all signed elementary products of A.

EXAMPLE

$$\det \begin{bmatrix} 1 & 2 \\ 3 & 4 \end{bmatrix} = (1 \times 4) - (2 \times 3) = -2$$

Problem Solving Examples:

Find the determinant of the following matrices:

(a) $[a_{11}]$ (b) $\begin{bmatrix} a_{11} & a_{12} \\ a_{21} & a_{22} \end{bmatrix}$ (c) $\begin{bmatrix} 0 & 0 & 0 \\ 0 & 0 & 0 \\ 0 & 0 & 0 \end{bmatrix}$

(a) $\det(A) = a_{11}$

(b) $\det(A) = \begin{bmatrix} a_{11} & a_{12} \\ a_{21} & a_{22} \end{bmatrix} = a_{11}a_{22} - a_{21}a_{12}$

(c) $\det(A) = [0] = 0$

2.2 Determinants by Row Reduction

THEOREM

A square matrix containing a row of zeros has a determinant of zero.

A square matrix is in upper triangular form if it has all zero entries below the main diagonal; it is in lower triangular form if it has all zero entries above the main diagonal; and it is in triangular form if it is in either upper or lower triangular form.

EXAMPLES

a)
$$\begin{bmatrix} 1 & 2 & 3 & 4 \\ 0 & 5 & 6 & 7 \\ 0 & 0 & 8 & 9 \\ 0 & 0 & 0 & 10 \end{bmatrix}$$ upper triangular form

b)
$$\begin{bmatrix} 1 & 0 & 0 & 0 \\ 2 & 3 & 0 & 0 \\ 4 & 5 & 6 & 0 \\ 7 & 8 & 9 & 10 \end{bmatrix}$$ lower triangular form

THEOREM

If A is an $n \times n$ triangular matrix, then the det(A) is the product of the main diagonal entries.

If a square matrix has two proportional rows, then its determinant is zero.

EXAMPLE

$$A = \begin{bmatrix} 3 & 4 & 2 \\ 4 & 5 & 1 \\ 6 & 8 & 4 \end{bmatrix}$$

det(A) = 0 since the third row is twice the first row.

THEOREM

Given A is any $n \times n$ matrix,

a) If A^* is the result of multiplying one row (or column) of A by a constant k, the det(A^*) = $k$det(A).

b) If A^* is the result of switching two rows (or columns) of A, then det(A^*) = –det(A).

c) If A^* is the result of adding a multiple of a row (or column) of A to another row (or column) of A, then $\det(A^*) = \det(A)$.

EXAMPLE

The determinant of the reduced row-echelon form of a matrix is equal to the determinant of the matrix.

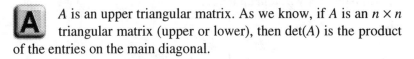
Problem Solving Examples:

Find the determinant of the matrix A where:

$$A = \begin{bmatrix} 2 & 7 & -3 & 8 & 3 \\ 0 & -3 & 7 & 5 & 1 \\ 0 & 0 & 6 & 7 & 6 \\ 0 & 0 & 0 & 9 & 8 \\ 0 & 0 & 0 & 0 & 4 \end{bmatrix}.$$

A is an upper triangular matrix. As we know, if A is an $n \times n$ triangular matrix (upper or lower), then $\det(A)$ is the product of the entries on the main diagonal.

Hence,

$$\det(A) = (2) \times (-3) \times (6) \times (9) \times (4) = -1{,}296.$$

Evaluate $\det(A)$ where:

$$A = \begin{bmatrix} 0 & 1 & 5 \\ 3 & -6 & 9 \\ 2 & 6 & 1 \end{bmatrix}$$

Interchange the first and second rows of matrix A, obtaining matrix

$$B = \begin{bmatrix} 3 & -6 & 9 \\ 0 & 1 & 5 \\ 2 & 6 & 1 \end{bmatrix}$$

and by the properties of determinants:

$$\det(A) = -\det(B).$$

$$= -\det \begin{bmatrix} 3 & -6 & 9 \\ 0 & 1 & 5 \\ 2 & 6 & 1 \end{bmatrix}$$

or

$$\det(A) = -3\det \begin{bmatrix} 1 & -2 & 3 \\ 0 & 1 & 5 \\ 2 & 6 & 1 \end{bmatrix}.$$

A common factor of 3 from the first row of the matrix B was taken out. Add -2 times the first row to the third row. The value of the determinant of A will remain the same.

Thus,

$$\det(A) = -3\det \begin{bmatrix} 1 & -2 & 3 \\ 0 & 1 & 5 \\ 0 & 10 & -5 \end{bmatrix}.$$

Add -10 times the second row to the third row.

Thus,

$$\det(A) = -3\det \begin{bmatrix} 1 & -2 & 3 \\ 0 & 1 & 5 \\ 0 & 0 & -55 \end{bmatrix}.$$

As we know, the determinant of a triangular matrix is equal to the product of the diagonal elements.

Thus,

$$\det(A) = (-3) \times (1) \times (1) \times (-55) = 165.$$

Compute the determinant of:

$$A = \begin{bmatrix} 1 & 0 & 0 & 3 \\ 2 & 7 & 0 & 6 \\ 0 & 6 & 3 & 0 \\ 7 & 3 & 1 & -5 \end{bmatrix}.$$

 Add -3 times the first column to the fourth column. But the value of the determinant A will not be changed.

$$A = \begin{bmatrix} 1 & 0 & 0 & 0 \\ 2 & 7 & 0 & 0 \\ 0 & 6 & 3 & 0 \\ 7 & 3 & 1 & -26 \end{bmatrix}$$

Now the above matrix A is a lower triangular matrix. Thus,

$$\det(A) = (1) \times (3) \times (-26) = -546.$$

2.3 Determinant Properties

If A is an $m \times n$ matrix, then the transpose of A, denoted by (A^t) is defined as the $n \times m$ matrix, where the rows and columns of A are switched.

EXAMPLE

If $A = \begin{bmatrix} 1 & 2 \\ 2 & 7 \\ 5 & 23 \end{bmatrix}$, then $A^t = \begin{bmatrix} 1 & 2 & 5 \\ 2 & 7 & 23 \end{bmatrix}$.

Properties of the transpose operation:

a) $(A^t)^t = A$

b) $(A + B)^t = A^t + B^t$

c) $(kA)^t = kA^t$ (where k is a scalar)

d) $(AB)^t = B^t A^t$

THEOREM

If A is a square matrix, then $\det(A) = \det(A^t)$. (Because of this theorem, all determinant theorems concerning the rows of a matrix also apply to the columns of a matrix.)

If A and B are square matrices of the same size, and k is a scalar, then:

a) $\det(kA) = k^n \det(A)$ (n is the number of rows of A)

b) $\det(AB) = \det(A)\det(B)$

THEOREM

A square matrix A is invertible if and only if $\det(A) \neq 0$.

If A is invertible, then $\det(A^{-1}) = \dfrac{1}{\det(A)}$.

Problem Solving Examples:

 Find the transpose of the following matrices:

$$A = \begin{bmatrix} 4 & -2 & 3 \\ 0 & 5 & -2 \end{bmatrix} \qquad B = \begin{bmatrix} 6 & 2 & -4 \\ 3 & -1 & 2 \\ 0 & 4 & 3 \end{bmatrix}$$

$$C = \begin{bmatrix} 5 & 4 \\ -3 & 2 \\ 2 & -3 \end{bmatrix} \qquad D = \begin{bmatrix} 3 & -5 & 1 \end{bmatrix} \qquad E = \begin{bmatrix} 2 \\ -1 \\ 3 \end{bmatrix}$$

 The transpose of A is obtained by interchanging the rows and columns of A.

$$A = \begin{bmatrix} 4 & -2 & 3 \\ 0 & 5 & -2 \end{bmatrix}$$

Then,

$$A^t = \begin{bmatrix} 4 & 0 \\ -2 & 5 \\ 3 & -2 \end{bmatrix}$$

$$B = \begin{bmatrix} 6 & 2 & -4 \\ 3 & -1 & 2 \\ 0 & 4 & 3 \end{bmatrix}$$

$$B^t = \begin{bmatrix} 6 & 3 & 0 \\ 2 & -1 & 4 \\ -4 & 2 & 3 \end{bmatrix}$$

$$C = \begin{bmatrix} 5 & 4 \\ -3 & 2 \\ 2 & -3 \end{bmatrix}; \text{ thus, } C^t = \begin{bmatrix} 5 & -3 & 2 \\ 4 & 2 & -3 \end{bmatrix}.$$

$$D = \begin{bmatrix} 3 & -5 & 1 \end{bmatrix}.$$

Then,

$$D^t = \begin{bmatrix} 3 \\ -5 \\ 1 \end{bmatrix}.$$

$$E = \begin{bmatrix} 2 \\ -1 \\ 3 \end{bmatrix}; \quad \text{hence, } E^t = \begin{bmatrix} 2 & -1 & 3 \end{bmatrix}.$$

 Let $A = \begin{bmatrix} 1 & 2 & 0 \\ 3 & -1 & 4 \end{bmatrix}$. Find (a) AA^t, (b) A^tA.

 Observe that if A is an $m \times n$ matrix, then A^t is an $n \times m$ matrix. Hence, the products AA^t and A^tA are always defined.

$$A = \begin{bmatrix} 1 & 2 & 0 \\ 3 & -1 & 4 \end{bmatrix}$$

then,

$$A^t = \begin{bmatrix} 1 & 3 \\ 2 & -1 \\ 0 & 4 \end{bmatrix}.$$

(a) $AA^t = \begin{bmatrix} 1 & 2 & 0 \\ 3 & -1 & 4 \end{bmatrix} \begin{bmatrix} 1 & 3 \\ 2 & -1 \\ 0 & 4 \end{bmatrix}$

$$= \begin{bmatrix} 1 \times 1 + 2 \times 2 + 0 \times 0 & 1 \times 3 + 2 \times (-1) + 0 \times 4 \\ 3 \times 1 + (-1) \times 2 + 4 \times 0 & 3 \times 3 + (-1) \times (-1) + 4 \times 4 \end{bmatrix}$$

$$= \begin{bmatrix} 1 + 4 + 0 & 3 - 2 + 0 \\ 3 - 2 + 0 & 9 + 1 + 16 \end{bmatrix} = \begin{bmatrix} 5 & 1 \\ 1 & 26 \end{bmatrix}.$$

(b) $A^t A = \begin{bmatrix} 1 & 3 \\ 2 & -1 \\ 0 & 4 \end{bmatrix} \begin{bmatrix} 1 & 2 & 0 \\ 3 & -1 & 4 \end{bmatrix}$

$$= \begin{bmatrix} 1 \times 1 + 3 \times 3 & 1 \times 2 + 3 \times (-1) & 1 \times 0 + 3 \times 4 \\ 2 \times 1 + (-1) \times 3 & 2 \times 2 + (-1) \times (-1) & 2 \times 0 + (-1) \times 4 \\ 0 \times 1 + 4 \times 3 & 0 \times 2 + 4 \times (-1) & 0 \times 0 + 4 \times 4 \end{bmatrix}$$

$$= \begin{bmatrix} 1 + 9 & 2 - 3 & 0 + 12 \\ 2 - 3 & 4 + 1 & 0 - 4 \\ 0 + 12 & 0 - 4 & 0 + 16 \end{bmatrix}$$

$$= \begin{bmatrix} 10 & -1 & 12 \\ -1 & 5 & -4 \\ 12 & -4 & 16 \end{bmatrix}$$

2.4 Cofactor Expansion and Cramer's Rule

If A is a square matrix, then the minor of entry a_{ij}, denoted M_{ij}, is defined to be the determinant of the submatrix remaining after the ith row and jth column of A are removed.

EXAMPLE

$$A = \begin{bmatrix} 7 & 1 & 3 \\ 1 & 3 & 5 \\ 17 & 4 & 20 \end{bmatrix}, \quad M_{11} = \det \begin{bmatrix} 3 & 5 \\ 4 & 20 \end{bmatrix} = 40$$

If A is a square matrix, then the cofactor of entry a_{ij}, denoted c_{ij}, is defined to be the scalar $(-1)^{i+j} M_{ij}$.

EXAMPLE

$$A = \begin{bmatrix} 7 & 1 & 3 \\ 1 & 3 & 5 \\ 17 & 4 & 20 \end{bmatrix}, \quad c_{11} = (-1)^{1+1} M_{11} = (-1)^2 \, 40 = 40$$

$c_{ij} = \pm M_{ij}$, depending on the position of the entry in relation to the matrix:

$$\begin{bmatrix} + & - & + & - & \dots \\ - & + & - & + & \dots \\ + & - & + & - & \dots \\ - & + & - & + & \dots \\ \vdots & \vdots & \vdots & \vdots & \end{bmatrix}.$$

EXAMPLE

$$c_{11} = +M_{11}, c_{12} = -M_{12}, c_{43} = -M_{43}, \text{ etc.}$$

If A is a square matrix, $\det(A)$ can be found by cofactor expansion along the ith row or jth column of A. This is done by multiplying the entries in the ith row of the jth column of A by their cofactors and summing the resulting products. Thus,

$$\det(A) = a_{i1} c_{i1} + a_{i2} c_{i2} + \dots$$

$$\text{or}$$

$$\det(A) = a_{1j} c_{2j} + a_{2j} c_{2j} + \dots$$

EXAMPLE

Expansion along the first row:

$$\det \begin{bmatrix} 2 & 4 & 3 \\ 7 & 2 & 12 \\ 1 & 3 & 9 \end{bmatrix} = (2)(-18) + 4(-51) + (3)(19) = -183.$$

If A is a square matrix and c_{ij} is the cofactor of a_{ij}, then the matrix of cofactors from A is:

$$\begin{bmatrix} c_{11} & c_{12} & \cdots \\ c_{21} & c_{22} & \cdots \\ \vdots & \vdots & \end{bmatrix}.$$

The transpose of this matrix is called the adjoint of A (adj(A)).

THEOREM

If A is an invertible matrix, then:

$$A^{-1} = \frac{1}{\det(A)} \, \text{adj}(A).$$

2.4.1 Cramer's Rule

If $AX = B$ is a system of n linear equations having unknowns x_1, $x_2, \ldots, x_n$, then the unique solution of the system is:

$$x_1 = \frac{\det(A_1)}{\det(A)}, \quad x_2 = \frac{\det(A_2)}{\det(A)}, \quad \ldots, x_n = \frac{\det(A_n)}{\det(A)}$$

where A_n is the matrix obtained by replacing the jth column of A with the column of constants of the system,

$$B = \begin{bmatrix} b_1 \\ b_2 \\ \vdots \\ b_n \end{bmatrix}.$$

Problem Solving Examples:

Find the determinant of the following matrix:

$$A = \begin{bmatrix} 2 & 0 & 3 & 0 \\ 2 & 1 & 1 & 2 \\ 3 & -1 & 1 & -2 \\ 2 & 1 & -2 & 1 \end{bmatrix}$$

Use cofactor expansion.

$$A = \begin{bmatrix} 2 & 0 & 3 & 0 \\ 2 & 1 & 1 & 2 \\ 3 & -1 & 1 & -2 \\ 2 & 1 & -2 & 1 \end{bmatrix}$$

Expanding along the first row:

$$\det (A) = 2 \begin{bmatrix} 1 & 1 & 2 \\ -1 & 1 & -2 \\ 1 & -2 & 1 \end{bmatrix} + 3 \begin{bmatrix} 2 & 1 & 2 \\ 3 & -1 & -2 \\ 2 & 1 & 1 \end{bmatrix}$$

Note that the minors, whose multiplying factors were zero, have been eliminated. This illustrates the general principle that, when evaluating determinants, expansion along the row (or column) containing the most zeros is the optimal procedure.

Add the second row to the first row for each of the 3×3 determinants:

$$\det A = 2 \begin{bmatrix} 0 & 2 & 0 \\ -1 & 1 & -2 \\ 1 & -2 & 1 \end{bmatrix} + 3 \begin{bmatrix} 5 & 0 & 0 \\ 3 & -1 & -2 \\ 2 & 1 & 1 \end{bmatrix}$$

Now expand the above determinants by minors using the first row.

$$\det A = 2(-2)\begin{bmatrix} -1 & -2 \\ 1 & 1 \end{bmatrix} + 3(5)\begin{bmatrix} -1 & -2 \\ 1 & 1 \end{bmatrix}$$
$$= (-4)(-1+2) + 15(-1+2)$$
$$= -4 + 15 = 11$$

a)   Given:

$$A = \begin{bmatrix} 5 & 2 & -1 \\ 3 & 1 & 2 \\ 2 & 7 & 4 \end{bmatrix}, \text{ find } \det(A).$$

b)   If

$$A = \begin{bmatrix} 2 & 3 & 1 \\ -2 & 4 & 5 \\ 2 & 0 & 7 \end{bmatrix}, \text{ find adj}(A).$$

The cofactors along the first column are:

$$c_{11} = (-1)^{1+1}\begin{bmatrix} 1 & 2 \\ 7 & 4 \end{bmatrix} = +(4-14) = -10$$

$$c_{21} = (-1)^{2+1}\begin{bmatrix} 2 & -1 \\ 7 & 4 \end{bmatrix} = -(8+7) = -15$$

$$c_{31} = (-1)^{3+1}\begin{bmatrix} 2 & -1 \\ 1 & 2 \end{bmatrix} = +(4+1) = 5$$

Hence,

$$= 5(-10) + 3(-15) + 2(5)$$
$$= -50 - 45 + 10$$
$$= -85$$

The adjoint is useful in finding the inverse of a non-singular matrix. The cofactors of A are:

$$c_{11} = + \begin{bmatrix} 4 & 5 \\ 0 & 7 \end{bmatrix} = (28 - 0) = 28$$

$$c_{12} = - \begin{bmatrix} -2 & 5 \\ 2 & 7 \end{bmatrix} = -(-14 - 10) = 24$$

$$c_{13} = + \begin{bmatrix} -2 & 4 \\ 2 & 0 \end{bmatrix} = (0 - 8) = -8$$

$$c_{21} = - \begin{bmatrix} 3 & 1 \\ 0 & 7 \end{bmatrix} = -(21 - 0) = -21$$

$$c_{22} = + \begin{bmatrix} 2 & 1 \\ 2 & 7 \end{bmatrix} = (14 - 2) = 12$$

$$c_{23} = - \begin{bmatrix} 2 & 3 \\ 2 & 0 \end{bmatrix} = -(0 - 6) = 6$$

$$c_{31} = + \begin{bmatrix} 3 & 1 \\ 4 & 5 \end{bmatrix} = (15 - 4) = 11$$

$$c_{32} = - \begin{bmatrix} 2 & 1 \\ -2 & 5 \end{bmatrix} = -(10 + 2) = -12$$

$$c_{33} = + \begin{bmatrix} 2 & 3 \\ -2 & 4 \end{bmatrix} = (8 + 6) = 14$$

The matrix of cofactors is:

$$\begin{bmatrix} 28 & 24 & -8 \\ -21 & 12 & 6 \\ 11 & -12 & 14 \end{bmatrix}$$

and the adjoint of A is:

$$\text{adj}(A) = \begin{bmatrix} 28 & -21 & 11 \\ 24 & 12 & -12 \\ -8 & 6 & 14 \end{bmatrix}.$$

Q A rancher sold 25 hogs and 60 sheep to Mr. Kay for $3,450. At the same prices, he sold 35 hogs and 50 sheep to Mr. Bea for $3,300. Find the price of each hog and sheep.

A Let y be the price of one sheep and x be the price of one hog. Since the rancher sold 25 hogs and 60 sheep for $3,450, then $25x + 60y = 3,450$. Similarly, $35x + 50y = 3,300$. By Cramer's Rule our linear system:

$$25x + 60y = 3,450$$
$$35x + 50y = 3,300$$

can be easily solved.

$$x = \frac{\begin{vmatrix} 3,450 & 60 \\ 3,300 & 50 \end{vmatrix}}{\begin{vmatrix} 25 & 60 \\ 35 & 50 \end{vmatrix}} \quad \text{and} \quad y = \frac{\begin{vmatrix} 25 & 3,450 \\ 35 & 3,300 \end{vmatrix}}{\begin{vmatrix} 25 & 60 \\ 35 & 50 \end{vmatrix}}$$

Thus,

$$x = \frac{(3,450)(50) - (3,300)(60)}{(25)(50) - (35)(60)}, \quad y = \frac{(25)(3,300) - (35)(3,450)}{(25)(50) - (35)(60)}$$

$$x = 30 \qquad\qquad y = 45$$

Therefore, the rancher sells hogs for $30 and sheep for $45.

Q Ms. Wong has a total of $4,200 invested in securities A, B, and C. The rates of annual dividends are 4%, 6%, and 5%, respectively, yielding total annual dividends of $214. If the sum of A and B is twice C, find the amount invested in each security.

A From the given conditions, we can form the system of linear equations:

$$A + B + C = 4,200$$
$$.04A + .06B + .05C = 214$$
$$A + B - 2C = 0$$

By multiplying through by 100, the second equation is equivalent to $4A + 6B + 5C = 21{,}400$. We can rewrite the equations above in matrix form:

$$\begin{bmatrix} 1 & 1 & 1 \\ 4 & 6 & 5 \\ 1 & 1 & -2 \end{bmatrix} \begin{bmatrix} A \\ B \\ C \end{bmatrix} = \begin{bmatrix} 4{,}200 \\ 21{,}400 \\ 0 \end{bmatrix}.$$

We can solve for A, B, and C by Cramer's Rule.

Thus,

$$A = \frac{\begin{vmatrix} 4{,}200 & 1 & 1 \\ 21{,}400 & 6 & 5 \\ 0 & 1 & -2 \end{vmatrix}}{\begin{vmatrix} 1 & 1 & 1 \\ 4 & 6 & 5 \\ 1 & 1 & -2 \end{vmatrix}}$$

$$= \frac{4{,}200(6)(-2) + 1(5)(0) + 1(21{,}400)(1) - 1(6)(0) - 1(21{,}400)(-2) - 4{,}200(5)(1)}{1(6)(-2) + 1(5)(1) + 1(4)(1) - 1(6)(1) - 1(4)(-2) - 1(5)(1)}$$

$$= \frac{-7{,}200}{-6}$$

$$= 1{,}200$$

$$B = \frac{\begin{vmatrix} 1 & 4{,}200 & 1 \\ 4 & 21{,}400 & 5 \\ 1 & 0 & 2 \end{vmatrix}}{\begin{vmatrix} 1 & 1 & 1 \\ 4 & 6 & 5 \\ 1 & 1 & -2 \end{vmatrix}}$$

$$= \frac{1(21{,}400)(2) + 4{,}200(5)(1) + 1(4)(0) - 1(21{,}400)(1) - 4{,}200(4)(2) - 1(5)(0)}{-6}$$

$$= \frac{-9{,}600}{-6}$$

$$= 1{,}600$$

$$C = \frac{\begin{vmatrix} 1 & 1 & 4,200 \\ 4 & 6 & 21,400 \\ 1 & 1 & 0 \end{vmatrix}}{\begin{vmatrix} 1 & 1 & 1 \\ 4 & 6 & 5 \\ 1 & 1 & -2 \end{vmatrix}}$$

$$= \frac{1(6)(0) + 1(21,400)(1) + 4,200(4)(1) - 4,200(6)(1) - 1(4)(0) - 1(21,400)(1)}{-6}$$

$$= \frac{-8,400}{-6}$$

$$= 1,400$$

Thus, Ms. Wong has \$1,200 invested in security A, \$1,600 invested in security B, and \$1,400 invested in security C.

Solve the following linear equations by using Cramer's Rule:

$$-2x_1 + 3x_2 - x_3 = 1$$
$$x_1 + 2x_2 - x_3 = 4$$
$$-2x_1 - x_2 + x_3 = -3$$

$$\det(A) = \begin{vmatrix} -2 & 3 & -1 \\ 1 & 2 & -1 \\ -2 & -1 & 1 \end{vmatrix}$$

$$\det(A) = -2\begin{vmatrix} 2 & -1 \\ -1 & 1 \end{vmatrix} - 3\begin{vmatrix} 1 & -1 \\ -2 & 1 \end{vmatrix} - 1\begin{vmatrix} 1 & 2 \\ -2 & -1 \end{vmatrix}$$
$$= -2(2-1) - 3(1-2) - (-1+4)$$
$$= -2 + 3 - 3 = -2$$

Since $\det(A) \neq 0$, the system has a unique solution. Now,

$$x_1 = \frac{\det(A_1)}{\det(A)}, \quad x_2 = \frac{\det(A_2)}{\det(A)}, \quad x_3 = \frac{\det(A_3)}{\det(A)}.$$

$\det(A_1)$ is the determinant of the matrix obtained by replacing the first column of A by the column of B. Thus,

$$\det(A_1) = \begin{bmatrix} 1 & 3 & -1 \\ 4 & 2 & -1 \\ 3 & -1 & 1 \end{bmatrix}$$

$$= -4$$

Then,

$$x_1 = \frac{-4}{-2} = 2.$$

$$x_2 = \frac{\begin{vmatrix} -2 & 1 & -1 \\ 1 & 4 & -1 \\ -2 & -3 & 1 \end{vmatrix}}{|A|}$$

$$= \frac{-6}{-2} = 3.$$

$$x_3 = \frac{\det(A_3)}{\det(A)} = \frac{\begin{vmatrix} -2 & 3 & 1 \\ 1 & 2 & 4 \\ -2 & -1 & -3 \end{vmatrix}}{-2}$$

$$= \frac{-8}{-2} = 4.$$

Thus,

$$x_1 = 2, \ x_2 = 3, \ x_3 = 4$$

is the unique solution to the given system.

 Solve the system of linear equations:

$$3x + 2y + 4z = 1$$
$$2x - y + z = 0$$
$$x + 2y + 3z = 1$$

Use Cramer's Rule to solve this system.

$$\begin{bmatrix} 3 & 2 & 4 \\ 2 & -1 & 1 \\ 1 & 2 & 3 \end{bmatrix} \begin{bmatrix} x \\ y \\ z \end{bmatrix} = \begin{bmatrix} 1 \\ 0 \\ 1 \end{bmatrix}$$

Then,

$$A = \begin{bmatrix} 3 & 2 & 4 \\ 2 & -1 & 1 \\ 1 & 2 & 3 \end{bmatrix}, \quad B = \begin{bmatrix} 1 \\ 0 \\ 1 \end{bmatrix}.$$

First, check that $\det(A) \neq 0$.

$$\det(A) = \begin{vmatrix} 3 & 2 & 4 \\ 2 & -1 & 1 \\ 1 & 2 & 3 \end{vmatrix}$$

$$\det(A) = 3 \begin{vmatrix} -1 & 1 \\ 2 & 3 \end{vmatrix} - 2 \begin{vmatrix} 2 & 1 \\ 1 & 3 \end{vmatrix} + 4 \begin{vmatrix} 2 & -1 \\ 1 & 2 \end{vmatrix}$$

$$= 3(-3 - 2) - 2(6 - 1) + 4(4 + 1)$$

$$= -15 - 10 + 20 = -5$$

Since $\det(A) \neq 0$, the system has a unique solution. Then,

$$x = \frac{\det(A_1)}{\det(A)}, \quad y = \frac{\det(A_2)}{\det(A)}, \quad z = \frac{\det(A_3)}{\det(A)}.$$

$\det(A_1)$ is the determinant of the matrix obtained by replacing the first column of A by the column vector B.

Thus,

$$\det(A_1) = \begin{vmatrix} 1 & 2 & 4 \\ 0 & -1 & 1 \\ 1 & 2 & 3 \end{vmatrix}.$$

Expand the determinant by minors, using the first column.

$$\det(A_1) = 1 \begin{vmatrix} -1 & 1 \\ 2 & 3 \end{vmatrix} + 1 \begin{vmatrix} 2 & 4 \\ -1 & 1 \end{vmatrix}$$
$$= 1(-3 - 2) + 1(2 + 4)$$
$$= -5 + 6 = 1$$

Now, since we have $\det(A) = -5$,

$$x = \frac{\det(A_1)}{\det(A)} = \frac{1}{-5} = -\frac{1}{5}$$

$$y = \frac{\det(A_2)}{\det(A)} = \frac{\begin{vmatrix} 3 & 1 & 4 \\ 2 & 0 & 1 \\ 1 & 1 & 3 \end{vmatrix}}{-5}$$

Now, expand $\det(A_2)$ along the second row.

$$\begin{vmatrix} 3 & 1 & 4 \\ 2 & 0 & 1 \\ 1 & 1 & 3 \end{vmatrix} = -2 \begin{vmatrix} 1 & 4 \\ 1 & 3 \end{vmatrix} - 1 \begin{vmatrix} 3 & 1 \\ 1 & 1 \end{vmatrix}$$
$$= -2(3 - 4) - 1(3 - 1)$$
$$= 2 - 2 = 0$$
$$y = \frac{0}{-5}$$
$$= 0$$

$$z = \frac{\det(A_3)}{\det(A)} = \frac{\begin{vmatrix} 3 & 2 & 1 \\ 2 & -1 & 0 \\ 1 & 2 & 1 \end{vmatrix}}{-5}$$

Expand determinant A_3 by minors, using the third column.

$$\begin{vmatrix} 3 & 2 & 1 \\ 2 & -1 & 0 \\ 1 & 2 & 1 \end{vmatrix} = +1 \begin{vmatrix} 2 & -1 \\ 1 & 2 \end{vmatrix} + 1 \begin{vmatrix} 3 & 2 \\ 2 & -1 \end{vmatrix}$$

$$= (4+1) + (-3-4)$$
$$= 5 - 7 = -2$$

Then,

$$z = \frac{-2}{-5} = \frac{2}{5}.$$

Thus,

$$x = -\frac{1}{5}, y = 0, z = \frac{2}{5}.$$

Quiz: Determinants

1. Let $M = \begin{bmatrix} 1 & 2 \\ 3 & 9 \end{bmatrix}$. The determinant of the adjoint of M is

    (A) 9.   (D) 18.

    (B) 6.   (E) 3.

    (C) 27.

2. Which of the 3×3 matrices has the following cofactor expansion?

$$3 \begin{bmatrix} 1 & 4 \\ -1 & 2 \end{bmatrix} - 2 \begin{bmatrix} 0 & 1 \\ 2 & -1 \end{bmatrix} + 5 \begin{bmatrix} 1 & 3 \\ 2 & 4 \end{bmatrix}$$

(A) $\begin{bmatrix} 3 & -2 & 5 \\ 1 & 1 & 4 \\ 2 & -1 & 2 \end{bmatrix}$.   (B) $\begin{bmatrix} 3 & 2 & 5 \\ 1 & 0 & 1 \\ 3 & 2 & -1 \end{bmatrix}$.

(C) $\begin{bmatrix} 3 & 1 & 3 \\ 2 & 2 & 4 \\ 5 & 0 & 1 \end{bmatrix}$.

(D) $\begin{bmatrix} 3 & 2 & 5 \\ 1 & 1 & 4 \\ 2 & -1 & 2 \end{bmatrix}$.

(E) $\begin{bmatrix} -3 & -2 & 5 \\ 1 & 1 & 4 \\ 2 & -1 & 2 \end{bmatrix}$.

3. Which of the following matrices has both a right inverse and a left inverse?

(A) $\begin{bmatrix} 1 & 2 & 3 \\ 2 & 4 & -1 \\ 0 & 0 & 0 \end{bmatrix}$.

(D) $\begin{bmatrix} 1 & 2 & 2 \\ -1 & 1 & 1 \\ 2 & 1 & 1 \end{bmatrix}$.

(B) $\begin{bmatrix} 0 & -3 & 1 \\ 4 & 0 & 2 \\ 3 & 1 & 0 \end{bmatrix}$.

(E) $\begin{bmatrix} 3 & 2 & 0 \\ 2 & -1 & 0 \\ 4 & -2 & 0 \end{bmatrix}$.

(C) $\begin{bmatrix} 0 & 1 & 2 \\ 0 & 3 & -1 \\ 0 & 2 & 4 \end{bmatrix}$.

4. A system of linear equations $a_{ij}x_j$, $i = 1,\ldots, n$, $j = 1,\ldots, m$, has a unique solution if and only if

(A) det $(a_{ij}) = 1$.

(D) det $(a_{ij}) = 0$.

(B) det $(a_{ij}) \neq 1$.

(E) det $(a_{ij}) < 0$.

(C) det $(a_{ij}) \neq 0$.

5. The following ratio of two determinants

$$\frac{\begin{vmatrix} 2 & -1 & 1 \\ 0 & 1 & 4 \\ 1 & 3 & 0 \end{vmatrix}}{\begin{vmatrix} 2 & -1 & 3 \\ 0 & 1 & -1 \\ 1 & 3 & -2 \end{vmatrix}}$$

is equal to the value of z in which of the following systems of equations?

(A) $2x - y + z = 1$
$y + 4z = 4$
$x + 3y = 0.$

(D) $2x - y + 3z = 0$
$y - x = 0$
$x + 3y - 2z = 0.$

(B) $2x + z = 1$
$-x - z = 4$
$x + 4y = 0.$

(E) $2x - y + 3z = 1$
$y - z = 4$
$x + 3y - 2z = 0.$

(C) $x + 4y + z = 0$
$-x + y + 3z = 0$
$2x + z = 0.$

6. The result of multiplying the determinant of the following matrix by 2 and then by 3,

$$\begin{bmatrix} 1 & 3 & 2 \\ 0 & 2 & -1 \\ 4 & -3 & 2 \end{bmatrix}$$

is equal to the determinant of which of the following matrices?

(A) $\begin{bmatrix} 1 & 3 & 2 \\ 0 & -2 & -5 \\ 4 & -3 & 2 \end{bmatrix}.$

(D) $\begin{bmatrix} -4 & -12 & -8 \\ 0 & -8 & 4 \\ -16 & 12 & -8 \end{bmatrix}.$

(B) $\begin{bmatrix} 0 & 6 & -3 \\ 4 & -3 & 2 \\ 2 & 6 & 4 \end{bmatrix}.$

(E) $\begin{bmatrix} 1 & 0 & 4 \\ 3 & 2 & -3 \\ 2 & -1 & 2 \end{bmatrix}.$

(C) $\begin{bmatrix} -4 & 3 & 2 \\ 0 & -8 & -1 \\ 4 & -3 & -8 \end{bmatrix}.$

7. If the determinants | A | = 3 and | B | = 2, find | $2(AB)^{-1}$ | for 4×4 matrices A and B.

(A) $\dfrac{1}{3}$.

(D) $\dfrac{8}{3}$.

(B) $\dfrac{2}{3}$.

(E) 12.

(C) $\dfrac{4}{3}$.

8. Let $A = \begin{bmatrix} a_{11} & a_{12} \ldots a_{1n} \\ a_{21} & a_{22} \ldots a_{2n} \\ \vdots & \vdots \\ a_{n1} & a_{n2} \ldots a_{nn} \end{bmatrix}$ and $B = \begin{bmatrix} b_{11} & b_{12} \ldots b_{1n} \\ b_{21} & b_{22} \ldots b_{2n} \\ \vdots & \vdots \\ b_{n1} & b_{n2} \ldots b_{nn} \end{bmatrix}$.

Furthermore, let A and B be such that $a_{ij} = b_{ij}$ when $i \geq j$. What is det $(A - B)$?

(A) 0.

(D) $a_{ij} - a_{ji}$.

(B) -1.

(E) $[a_{ij}]^2$.

(C) $+1$.

9. If the determinant of the matrix:

$$\begin{bmatrix} 1 & 0 & 0 & 0 & 0 \\ x & 2 & 0 & 0 & 0 \\ x_2 & x_3 & 3 & 0 & 0 \\ x_3 & x_4 & x_5 & 4 & 0 \\ x_4 & x_5 & x_6 & x_7 & 0 \end{bmatrix}$$

is zero, how many values of x are possible?

(A) 0. (D) 3.

(B) 1. (E) ∞.

(C) 2.

10. Which of the following statements about the determinants of the matrices A and B is true?

(A) If B is obtained from A by interchanging two rows, $\det(B) - \det(A) = 0$.

(B) If B is obtained from A by multiplying one row of A by some scalar c, then $\det(B) = c \det(A)$.

(C) If any two rows in A are identical, $\det(A) = 1$.

(D) If B is obtained from A by multiplying one row and adding the result to another row, $\det(B) = -\det(A)$.

(E) The determinant of the identity matrix is equal to -1.

ANSWER KEY

1.	(E)	6.	(B)
2.	(D)	7.	(D)
3.	(B)	8.	(A)
4.	(C)	9.	(E)
5.	(E)	10.	(B)

CHAPTER 3

Vector Spaces

3.1 Euclidean *n*-Space

An ordered *n*-tuple is a sequence of *n* real numbers, where *n* is a positive integer. An ordered 2-tuple is called an ordered pair and an ordered 3-tuple is called an ordered triple.

EXAMPLE:

(4, 7, 12, 9, 3) is an ordered 5-tuple.

The *n*-space R^n is the set of all ordered *n*-tuples. R^1 is the set of all real numbers and can be written as R.

3.2 Vector Spaces

Let V be a set of objects on which two operations are defined, addition and multiplication by scalars (real numbers). Let it also be given that $\vec{u}$, $\vec{v}$, and $\vec{w}$ are members of V, and a and b are scalars. If V conforms to the following rules, then V is called a vector space, and $\vec{u}$, $\vec{v}$, and $\vec{w}$ are called vectors:

a) $\vec{u} + \vec{v} \in V$ (Closure under addition)

b) $\vec{u} + \vec{v} = \vec{v} + \vec{u}$ (Addition is commutative)

c) $\vec{u} + \left(\vec{v} + \vec{w} \right) = \left(\vec{u} + \vec{v} \right) + \vec{w}$ (Addition is associative)

d) The zero vector **0** is an element of V such that $0 + \vec{u} = \vec{u}$. (Existence and uniqueness of a zero element)

e) The negative of $\vec{u}$, $-\vec{u}$, is an element of V such that $\vec{u} + \left(-\vec{u} \right) = 0$. (Existence and uniqueness of an additive inverse)

f) $a\vec{u} \in V$ (Closure under scalar multiplication)

g) $a\left(\vec{u} + \vec{v} \right) = a\vec{u} + u\vec{v}$ (The first distributive law)

h) $(a + b)\vec{u} = a\vec{u} + b\vec{u}$ (The second distributive law)

i) $a\left(b\vec{u} \right)\vec{u} = (ab)\vec{u}$ (Associativity of scalar multiplication)

j) $1\vec{u} = \vec{u}$ (The existence of a unit element)

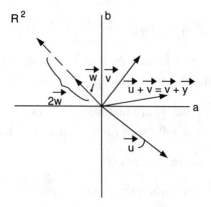

Figure 3.1 R^2 **is a vector space.**

To efficiently test if an object v is a vector, you must check to find if all of the following hold for v (for easy identification, all vectors are identified by arrows above the letters):

a) $\vec{0}\, v = \vec{0}$

b) $k\vec{0} = \vec{0}$

c) $(-1)\vec{v} = -\vec{v}$

d) If $k\vec{v} = \vec{0}$, then $k = 0$ or $\vec{v} = \vec{0}$.

A vector space can be in the form of n-space, with its vectors in the form of n-tuples. For this representation, the following conditions apply:

a) Two vectors $\vec{u} = (u_1, u_2, \ldots, u_n)$ and $\vec{v} = (v_1, v_2, \ldots, v_n)$ in R^n are equal if $u_1 = v_1$, $u_2 = v_2, \ldots, u_n = v_n$.

b) The sum $\vec{u} + \vec{v}$ is defined to be
$$\vec{u} + \vec{v} = (u_1 + v_1, u_2 + v_2, \ldots, u_n + v_n).$$

c) If L is scalar, then the scalar multiple is defined to be
$$L\vec{u} = (Lu_1, Lu_2, \ldots, Lu_n).$$

d) The zero vector is defined to be $\vec{0} = (0, 0, \ldots, 0)$.

e) The negative of $\vec{u}$ is defined to be
$$-\vec{u} = (-u_1, -u_2, \ldots, -u_n).$$

f) $\vec{u} - \vec{v} = \vec{u} + \left(-\vec{v}\right)$.

A vector space in R^1, R^2, or R^3 can be represented in a geometric form with a line, a plane, or two planes representing the vector space and either dots or lines representing the vectors.

EXAMPLES

a) $\vec{v} = (4)$ in R^1

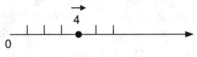

Figure 3.2

b) $\vec{u} = (6, 3)$ in R^2

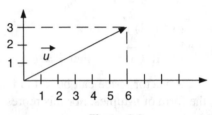

Figure 3.3

Problem Solving Examples:

Q Show that the space R^n (comprised of n-tuples of real numbers $(x_1,...x_n)$) is a vector space over the field R of real numbers. The operations are addition of n-tuples, i.e., $(x_1,x_2,...,x_n) + (y_1, y_2,..., y_n) = (x_1 + y_1, x_2 + y_2,..., x_n + y_n)$, and scalar multiplication, $a(x_1, x_2,..., x_n) = (ax_1, ax_2,..., ax_n)$ where $a \in R$.

A Any set that satisfies the axioms for a vector space over a field is known as a vector space. We must show that R^n satisfies the vector space axioms. The axioms fall into two distinct categories:

a)   the axioms of addition for elements of a set,

b)   the axioms involving multiplication of vectors by elements from the field.

1)   Closure under addition.

By definition,

$$(x_1, x_2,..., x_n) + (y_1, y_2,..., y_n) = (x_1 + y_1, x_2 + y_2,..., x_n + y_n).$$

Now, since $x_1, y_1, x_2, y_2,..., x_n, y_n$ are real numbers, the sums of $x_1 + y_1, x_2 + y_2,..., x_n + y_n$ are also real numbers. Therefore,

$$(x_1 + y_1, x_2 + y_2,\ldots, x_n + y_n)$$

is also an *n*-tuple of real numbers; hence, it belongs to R^n.

2) Addition is commutative.

The numbers $x_1, x_2,\ldots, x_n$ are the coordinates of the vector

$$(x_1, x_2,\ldots, x_n),$$

and $y_1, y_2,\ldots, y_n$ are the coordinates of the vector $(y_1, y_2,\ldots, y_n)$.

Show $(x_1, x_2,\ldots, x_n) + (y_1, y_2,\ldots, y_n) = (y_1,\ldots, y_n) + (x_1,\ldots, x_n)$. Now, the coordinates $x_1 + y_1, x_2 + y_2,\ldots, x_n + y_n$ are sums of real numbers. Since real numbers satisfy the commutativity axiom,

$$x_1 + y_1 = y_1 + x_1, x_2 + y_2 = y_2 + x_2,\ldots, x_n + y_n = y_n + x_n.$$

Thus,

$$(x_1, x_2,\ldots, x_n) + (y_1, y_2,\ldots, y_n)$$

$$= (x_1 + y_1, x_2 + y_2,\ldots, x_n + y_n) \quad \text{(by definition)}$$

$$= (y_1 + x_1, y_2 + x_2,\ldots, y_n + x_n) \quad \text{(by commutativity of real numbers)}$$

$$= (y_1, y_2,\ldots, y_n) + (x_1, x_2,\ldots, x_n) \quad \text{(by definition)}$$

We have shown that *n*-tuples of real numbers satisfy the commutativity axiom for a vector space.

3) Addition is associative.

$(a + b) + c = a + (b + c)$. Let $(x_1, x_2,\ldots, x_n)$, $(y_1, y_2,\ldots, y_n)$, and $(z_1, z_2,\ldots, z_n)$ be three points in R^n.

Now,

$$\left((x_1, x_2 \ldots, x_n) + (y_1, y_2,\ldots, y_n)\right) + (z_1, z_2,\ldots, z_n)$$

$$= (x_1 + y_1, x_2 + y_2,\ldots, x_n + y_n) + (z_1, z_2,\ldots, z_n) \qquad (1)$$

$$= (x_1 + y_1) + z_1, (x_2 + y_2) + z_2,\ldots, (x_n + y_n) + z_n$$

The coordinates $(x_i + y_i) + z_i$ ($i = 1,\ldots, n$) are real numbers. Since real numbers satisfy the associativity axiom, $(x_i + y_i) + z_i = x_i + (y_i + z_i)$. Hence, (1) may be rewritten as:

$$\left(x_1 + \left(y_1 + z_1\right), x_2 + \left(y_2 + z_2\right), \ldots, x_n + \left(y_n + z_n\right)\right)$$
$$= \left(x_1, x_2, \ldots, x_n\right) + \left(y_1 + z_1, y_2 + z_2, \ldots, y_n + z_n\right)$$
$$= \left(x_1, x_2, \ldots, x_n\right) + \left(\left(y_1, y_2, \ldots, y_n\right) + \left(z_1, z_2, \ldots, z_n\right)\right).$$

4) Existence and uniqueness of a zero element.

The set R^n should have a member $(a_1, a_2, \ldots, a_n)$ such that for any point $(x_1, \ldots, x_n)$ in R^n, $(x_1, x_2, \ldots, x_n) + (a_1, a_2, \ldots, a_n) = (x_1, x_2, \ldots, x_n)$. The point $(0, 0, 0, \ldots, 0)$ (n zeros), where 0 is the unique zero of the real number system, satisfies this requirement.

5) Existence and uniqueness of an additive inverse.

Let $(x_1, x_2, \ldots, x_n) \in R^n$. An additive inverse of $(x_1, x_2, \ldots, x_n)$ is an n-tuple $(a_1, a_2, \ldots, a_n)$ such that $(x_1, x_2, \ldots, x_n) + (a_1, a_2, \ldots, a_n) = (0, 0, \ldots, 0)$. Since $x_1, x_2, \ldots, x_n$ belong to the real number system, they have unique additive inverses $(-x_1), (-x_2), \ldots, (-x_n)$. Consider:

$$(-x_1, -x_2, \ldots, -x_n) \in R^n$$
$$(x_1, x_2, \ldots, x_n) + (-x_1, -x_2, \ldots, -x_n)$$
$$= \left(x_1 + (-x_1), x_2 + (-x_2), \ldots, x_n + (-x_n)\right)$$
$$= (0, 0, \ldots, 0)$$

We now turn to the axioms involving scalar multiplication.

6) Closure under scalar multiplication.

By definition, $a(x_1, x_2, \ldots, x_n) = (ax_1, ax_2, \ldots, ax_n)$ where the coordinates ax_i are real numbers. Hence, $(ax_1, ax_2, \ldots, ax_n) \in R^n$.

7) Associativity of scalar multiplication.

Let a, b be elements of R. We must show that $(ab)(x_1, x_2, \ldots, x_n) = a(bx_1, bx_2, \ldots, bx_n)$. But, since a, b, and $x_1, x_2, \ldots, x_n$ are real numbers, $(ab)(x_1, x_2, \ldots, x_n) = (abx_1, abx_2, \ldots, abx_n) = a(bx_1, bx_2, \ldots, bx_n)$.

8) The first distributive law.

We must show that $a\left(\vec{x} + \vec{y}\right) = a\vec{x} + a\vec{y}$ where $\vec{x}$ and $\vec{y}$ are vectors in R^n and $a \in R$.

$$a\big[(x_1, x_2, \ldots, x_n) + (y_1, y_2, \ldots, y_n)\big]$$
$$= a\big((x_1 + y_1), (x_2 + y_2), \ldots, (x_n + y_n)\big) \tag{2}$$
$$= \big(a(x_1 + y_1), a(x_2 + y_2), \ldots, a(x_n + y_n)\big)$$

(by definition of scalar multiplication).

Since each coordinate is a product of a real number and the sum of two real numbers, $a(x_i + y_i) = ax_i + ay_i$. Hence, (2) becomes

$$\big[(ax_1 + ay_1), (ax_2 + ay_2), \ldots, (ax_n + ay_n)\big]$$
$$= \big(ax_1, ax_2, \ldots, ax_n\big) + \big(ay_1, ay_2, \ldots, ay_n\big)$$
$$= ax + ay$$

9) The second distributive law.

We must show that $(a + b)\, x = ax + bx$ where $a, b \in R$ and x is a vector in R^n. Since $a + b$ is also a scalar,

$$(a + b)\,(x_1, x_2, \ldots, x_n)$$
$$= \big((a + b)\,x_1, (a + b)\,x_2, \ldots, (a + b)\,x_n\big) \tag{3}$$

Since a, b, and x_i are all real numbers, then $(a + b)x_i = ax_i + bx_i$. Therefore, (3) becomes:

$$\big((ax_1 + bx_1), (ax_2 + bx_2), \ldots, (ax_n + bx_n)\big)$$
$$= \big(ax_1, ax_2, \ldots, ax_n\big) + \big(bx_1, bx_2, \ldots, bx_n\big)$$
$$= a\big(x_1, x_2, \ldots, x_n\big) + b\big(x_1, x_2, \ldots, x_n\big)$$

10) The existence of a unit element from the field.

We require that there exist a scalar in the field R, call it "1," such that $1\,(x_1, x_2, \ldots, x_n) = (x_1, x_2, \ldots, x_n)$.

Now the real number 1 satisfies this requirement.

Since a set defined over a field that satisfies (1) – (10) is a vector space, the set R^n of n-tuples is a vector space when equipped with the given operations of addition and scalar multiplication.

Show that the set of all arithmetical progressions forms a two-dimensional vector space over the field of real numbers.

 An arithmetic progression of real numbers is a sequence of real numbers such that the difference between any two successive terms is a constant $\left(|x_{n+1} - x_n| = \text{constant}\right)$.

To make things clearer, consider a numerical example. Let:

$$x = (2, 5, 8, 11, 14,\ldots, 3n - 1,\ldots)$$

and

$$y = (6, 11, 16, 21, 26,\ldots, 5n + 1,\ldots).$$

Then $(x + y) = (8, 16, 24, 32, 40,\ldots, 8n,\ldots)$. Letting $a = 3$, we have $ax = 3(2, 5, 8, 11, 14,\ldots, 3n - 1,\ldots) = (6, 15, 24, 33, 42,\ldots, 9n - 3,\ldots)$.

It should be clear that the sum of two progressions is a progression, and that a progression multiplied by a scalar is a progression. That is, the set of progressions is closed with respect to addition and scalar multiplication.

If we view a progression as an infinite tuple of real numbers, we see that they satisfy the axioms for a vector field. Thus,

a) $\left(\vec{x} + \vec{y}\right) = \left(\vec{y} + \vec{x}\right)$;

b) $\left(\vec{x} + \vec{y}\right) + \vec{z} = \vec{x} + \left(\vec{y} + \vec{z}\right)$;

c) $\vec{x} = (x_1, x_2,\ldots, x_n,\ldots) + \vec{0} = (x_1, x_2,\ldots, x_n,\ldots),\ \vec{0} = (0, 0,\ldots)$;

d) $(x_1, x_2,\ldots, x_n,\ldots) + ((-x_1), (-x_2),\ldots,) = (0, 0,\ldots)$.

Similarly, the axioms of scalar multiplication also hold. Thus, the set of all arithmetic progressions is a vector space.

 Show that V, the set of all functions from a set $S \neq \varnothing$ to the field R, is a vector space under the following operations: if $f(s)$ and $g(s) \in V$, then $(f + g)(s) = f(s) + g(s)$. If c is a scalar from R, then $(cf)(s) = cf(s)$.

Since the points of V are functions, V is called a function space. Because our field is R, V is the space of real-valued functions defined on S. Also, since addition of real numbers is commutative, $f(s) + g(s) = g(s) + f(s)$. (Here $f(s)$ and $g(s)$ are real numbers, the values of the functions f and g at the point $s \in S$.) Since addition in R is associative, $f(s) + [g(s) + h(s)] = [f(s) + g(s)] + h(s)$ for all $s \in S$. Hence, addition of functions is associative. Next, the unique zero vector is the zero function which assigns to each $s \in S$ the value $0 \in R$. For each f in V, let $(-f)$ be the function given by

$$(-f)(s) = -f(s).$$

Then $f + (-f) = 0$ as required.

Next, we must verify the scalar axioms. Since multiplication in R is associative,

$$(cd) f(s) = c\, (df(s)).$$

The two distributive laws are: $c\,(f + g)\,(s) = cf(s) + cg(s)$ and $(c + d) f(s) = cf(s) + df(s)$. They hold by the properties of the real numbers. Finally, for $1 \in R$, $1f(s) = f(s)$.

Thus, V is a vector space. If, instead of R we had considered a field F, we would have had to verify the above axioms using the general properties of a field (any set that satisfies the axioms for a field).

Q Show that the set of semi-magic squares of order 3×3 form a vector space over the field of real numbers with addition defined as: $a_{ij} + b_{ij} = (a + b)_{ij}$ for $i, j = 1, \ldots, 3$.

A First, consider an example of a magic square.

$$\begin{bmatrix} 4 & 9 & 2 \\ 3 & 5 & 7 \\ 8 & 1 & 6 \end{bmatrix} \tag{1}$$

We see that in (1)

a)   every row has the same total, T;

b)   every column has the same total, T;

c)   the two diagonals have the same total, again T.

A square array of numbers that satisfies a) and b) but not c) is called a semi-magic square. For example, from (1)

$$\begin{bmatrix} 6 & 2 & 7 \\ 8 & 4 & 3 \\ 1 & 9 & 5 \end{bmatrix}$$

is semi-magic. Let M be the set of 3×3 semi-magic squares. We should notice that when two semi-magic squares are added, the result is a semi-magic square. For example, suppose m_1 and $m_2 \in M$ have row (and column) sums of T_1 and T_2, respectively; then each row and column in $m_1 + m_2$ will have a sum of $T_1 + T_2$.

Now, for m_1, m_2, and $m_3 \in M$,

a) $m_1 + m_2 = m_2 + m_1$.

This follows from the commutativity property of the real numbers.

b) $(m_1 + m_2) + m_3 = m_1 + (m_2 + m_3)$

Again, since the elements of m_1, m_2, and m_3, are real numbers and real numbers obey the associative law, semi-magic squares are associative with respect to addition.

c)   Existence of a zero element

$$\begin{bmatrix} a_{11} & a_{12} & a_{13} \\ a_{21} & a_{22} & a_{23} \\ a_{31} & a_{32} & a_{33} \end{bmatrix} + \begin{bmatrix} 0 & 0 & 0 \\ 0 & 0 & 0 \\ 0 & 0 & 0 \end{bmatrix} = \begin{bmatrix} a_{11} & a_{12} & a_{13} \\ a_{21} & a_{22} & a_{23} \\ a_{31} & a_{32} & a_{33} \end{bmatrix}$$

The 3×3 array with zeros everywhere is a semi-magic square.

d)   Existence of an additive inverse

If we replace every element in a semi-magic square with its negative, the result is a semi-magic square. Adding, we obtain the zero semi-magic square.

Next, let a be a scalar from the field of real numbers. Then, am_1 is still a semi-magic square. The two distributive laws hold, and $(ab)m_1 = a(bm_1)$ from the properties of the real numbers. Finally, $1(m_1) = m_1$, i.e., multiplication of a vector in M by the unit element from R leaves the vector unchanged.

Thus, M is a vector space.

3.3 Subspaces

Let V be a vector space, and let W be a subset of V. W is itself a vector space, and thus a subspace of V if W satisfies the following conditions:

a) If $\vec{v}$ and $\vec{w}$ are elements of W, then their sum $\vec{v} + \vec{w}$ is also an element of W.

b) If $\vec{v}$ is an element of W and c is a number, then $c\vec{v}$ is an element of W.

c) The element $\vec{0}$ of V is also an element of W.

Every vector space has at least two subspaces; the vector space itself, and the zero subspace (the space consisting of the zero vector only).

The vector consisting of a solution of a system of linear equations is called the solution vector of the system.

The set of all solution vectors is called the solution space of the system.

EXAMPLE

$$x_1 = 3$$
$$x_1 + x_2 = 5$$

solution vector $= (3, 2)$

A vector $\vec{w}$ is called a linear combination of the vectors $\vec{v}_1, \vec{v}_2 ..., \vec{v}_n$ if $\vec{w} = k_1 \vec{v}_1 + k_2 \vec{v}_2 + ... + k_n \vec{v}_n$, where $k_1, k_2, ..., k_n$ are scalars.

EXAMPLE

If $\vec{w} = (10, 24, 54)$ and $\vec{v}_1 = (1, 2, 3)$ and $\vec{v}_2 = (2, 5, 12)$, then $\vec{w}$ is a linear combination of $\vec{v}_1$ and $\vec{v}_2$ since

$$w = 2v_1 + 4v_2 = 2(1,2,3) + 4(2,5,12)$$
$$= (2,4,6) + (8,20,48)$$
$$= (10,24,54)$$

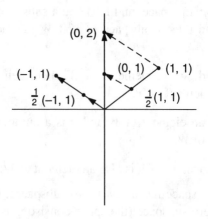

Figure 3.4 (0, 2) is a linear combination of (1, 1) and (–1, 1), and (0, 1) is also a linear combination of (1, 1) and (–1, 1). (0, 1) = $\frac{1}{2}$(1, 1) + $\frac{1}{2}$(–1, 1).

If every vector in a vector space V can be expressed as a linear combination of a subspace of V, then the vectors in that subspace are said to span V.

THEOREM

If $\vec{x}_1 \ \vec{x}_2, ..., \ \vec{x}_n$ are vectors in the vector space X, then:

    a)   the set Y of all linear combinations of $\vec{x}_1 \ \vec{x}_2, ..., \ \vec{x}_n$ is a subspace of X.

    b) Y is the smallest subset of X which contains $\vec{x}_1 \ \vec{x}_2, ..., \ \vec{x}_n$.

The vector space spanned by the set of vectors in a vector space V is called the linear space of V, denoted by $(\text{lin}(V))$.

Problem Solving Examples:

Q Give an example of a pair of non-collinear vectors in R^2. Then, show that the point $(x_1, x_2) = (8, 7)$ can be expressed as a linear combination of the non-collinear vectors.

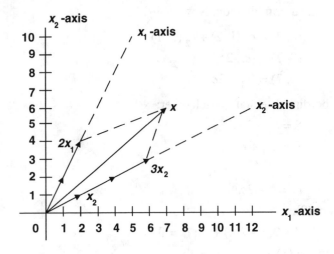

Figure 3.5

A Two vectors in R^2 are non-collinear when neither is a multiple of the other; that is $\vec{v_1}, \vec{v_2}$ ($\vec{v_1}$ and $\vec{v_2}$ are not both the $\vec{0}$ vector) are non-collinear if there is no scalar c such that $\vec{v_2} = c\,\vec{v_1}$. As justification, note that a vector in R^2 is an arrow (straight line) from the origin to some point (x_1, x_2). If we multiply this vector by a scalar, we will only expand or shorten the old vector to obtain a new vector.

Let $\vec{e_1} = (1, 2)$; $\vec{e_2} = (2, 1)$. Then e_1 and e_2 are non-collinear since setting

$$(2, 1) = c (1, 2)$$

we obtain

$$(2, 1) = (c, 2c)$$
$$2 = c$$
$$1 = 2c$$

There exists no c satisfying these two inconsistent equations.

Two non-collinear vectors in R^2 are sufficient to express any vector in R^2 as a linear combination. Now,

$$(8, 7) = c_1 (1, 2) + c_2 (2, 1)$$
$$(8, 7) = (c_1, 2c_1) + (2c_2, c_2)$$
$$(8, 7) = (c_1 + 2c_2, 2c_1 + c_2)$$

Setting coordinates equal to each other:

$$8 = c_1 + 2c_2 \tag{1}$$
$$7 = 2c_1 + c_2 \tag{2}$$

From (1),

$$c_1 = 8 - 2c.$$

Then, from (2),

$$7 = 2(8 - 2c_2) + c_2; c_2 = 3.$$

Substituting in (1),

$$8 = c_1 + 2(3)$$

gives $c_1 = 2$,

and

$$(8, 7) = c_1 (1, 2) + c_2 (2, 1)$$
$$= 2(1, 2) + 3 (2, 1)$$

 Are the following sets subspaces?

a) $W \subset R^3$ consists of all vectors of the form $(a, b, 1)$ where a and b are any real numbers.

b) $W \subset R^4$ contains all vectors of the form $(a, b, a - b, a + 2b)$ for $a, b \in R$.

a) To check whether W is a subspace, let $w_1 = (a_1, b_1, 1)$ and $w_2 = (a_2, b_2, 1)$ be vectors in W. Then,

$$w_1 + w_2 = (a_1, b_1, 1) + (a_2, b_2, 1) \qquad (1)$$
$$= (a_1 + a_2, b_1 + b_2, 2)$$

But $(a_1 + a_2, b_1 + b_2, 2)$ is not of the form $(a, b, 1)$ since its third component is not a 1. So $w_1 + w_2 \notin W$.

Hence, W is not closed under addition, i.e., W is not a subspace.

b) Let $w_1 = (a_1, b_1, a_1 - b_1, a_1 + 2b_1)$ and

$w_2 = (a_2, b_2, a_2 - b_2, a_2 + 2b_2)$.

Then, $w_1 + w_2$

$$= (a_1 + a_2, b_1 + b_2, a_1 + a_2 - (b_1 + b_2), a_1 + a_2 + 2(b_1 + b_2)) \qquad (2)$$

Vectors of the form (2) are again in W. To see this, replace $a_1 + a_2$ by a' and $b_1 + b_2$ by b'.

Thus, $w_1 + w_2 = (a', b', a' - b', a' + 2b')$.

Hence, W is closed under addition. Next, if k is a scalar, then

$$kw_1 = k(a_1, b_1, a_1 - b_1, a_1 + 2b_1) \qquad (3)$$
$$= [ka_1, kb_1, ka_1 - kb_1, ka_1 + 2kb_1]$$

Vectors of the form (3) are again in W. Thus, W is a subspace of R^4.

Let W be the set consisting of all 2×3 matrices of the form

$$\begin{bmatrix} a & b & 0 \\ 0 & c & d \end{bmatrix}$$ where $a, b, c,$ and d are real numbers. (1)

Show that W is a subspace of V, the set of all 2×3 matrices under the operation of addition over the field of real numbers.

To show that $W \subseteq V$ is a subspace we must show that, for $w_1, w_2 \in W$ and $k \in R$ (the field),

a) $W \neq \varnothing$

b) $w_1 + w_2 \in W$

c) $kw_1 \in W$

To show a), let $a = b = c = d = 0$ in the matrix (1). Then W contains at least the zero element and is non-empty.

To show b), let

$$w_1 = \begin{bmatrix} a_1 & b_1 & 0 \\ 0 & c_1 & d_1 \end{bmatrix} \text{ and } w_2 = \begin{bmatrix} a_2 & b_2 & 0 \\ 0 & c_2 & d_2 \end{bmatrix}$$

be two matrices in W. Then, by the rules of matrix addition,

$$w_1 + w_2 = \begin{bmatrix} a_1 + a_2 & b_1 + b_2 & 0 \\ 0 & c_1 + c_2 & d_1 + d_2 \end{bmatrix}.$$

But, by the rules of addition for real numbers, $a_1 + a_2, b_1 + b_2, c_1 + c_2$, and $d_1 + d_2$ are real numbers. Hence, $w_1 + w_2 \in W$.

Finally, to show c) let

$$kw_1 = k \begin{bmatrix} a_1 & b_1 & 0 \\ 0 & c_1 & d_1 \end{bmatrix}. \tag{2}$$

By the rules of scalar multiplication for matrices, (2) becomes

$$\begin{bmatrix} ka_1 & kb_1 & 0 \\ 0 & kc_1 & kd_1 \end{bmatrix}.$$

Since k is a real number, by the rules for multiplication of real numbers: ka_1, kb_1, kc_1, and kd_1 are real numbers.

Hence, $kw_1 \in W$.

Since W satisfies a), b), and c), we conclude that it is a subspace of $V_{2 \times 3}$.

Let $V_{n \times n}$ be the vector space of all square $n \times n$ matrices over the field of real numbers. Let W consist of all matrices which commute with a given matrix T, i.e., $W = \{A \in V: AT = TA\}$. Show that W is a subspace of V.

We must show that a) $W \neq \varnothing$; b) for $w_1, w_2 \in W$ and $a, b \in R$ (the field), $(aw_1 + bw_2) \in W$.

Note that this is a way of combining the requirements of closure for addition and scalar multiplication.

To show a), we note that for given T, if A is the zero matrix of order n, then $AT = TA$. Hence, W contains the zero vector and is non-empty.

To show b), we must show that $(aw_1 + bw_2)T = T(aw_1 + bw_2)$. Now, by the right distributive law,

$$(aw_1 + bw_2)T = (aw_1)T + (bw_2)T \qquad (1)$$
$$= a(w_1 T) + b(w_2 T) \qquad (2)$$

Since w_1 and $w_2 \in W$, the set of elements that commute with T, (2) becomes

$$a(Tw_1) + b(Tw_2). \qquad (3)$$

Now we get

$$T(aw_1) + T(bw_2), \qquad (4)$$

since any scalar commutes with a vector in $V_{n \times n}$. Finally, (4) gives

$$T(aw_1 + bw_2) \qquad (5)$$

by the left distributive law. (1) – (5) show that

$$(aw_1 + bw_2)T = T(aw_1 + bw_2), \text{ so } aw_1 + bw_2 \in W.$$

Therefore, a) and b) imply W is a vector space.

3.4   Linear Independence

Let V be a vector space, and let $\vec{v_1}, \ldots, \vec{v_n}$ be elements of V. Vectors $\vec{v_1}, \ldots, \vec{v_n}$ are linearly dependent if there exists numbers $a_1, \ldots, a_n$ not all equal to zero such that,

$$a_1 \vec{v_1} + \ldots + a_n \vec{v_n} = 0.$$

If such numbers do not exist, then $\vec{v}_1,\ldots,\vec{v}_n$ are linearly independent.

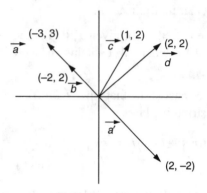

Figure 3.6

$\vec{a}$ and $\vec{a'}$ are linearly dependent since $-1\,\vec{a} = \vec{a'}$. $\vec{a}$, $\vec{b}$, $\vec{c}$ or $\vec{a'}$, $\vec{b}$, $\vec{c}$ are a linearly dependent set of vectors since $\vec{b} = \frac{1}{6}\vec{a} + \frac{3}{4}\vec{c}$.

If $Z = \left\{ \vec{z}_1, \vec{z}_2, \ldots, \vec{z}_r \right\}$ is a set of vectors in R^n, and $r > n$, then Z is linearly dependent.

Problem Solving Examples:

Show that $p_1(x) = x + 2x^2$, $p_2(x) = 1 + 2x + x^2$, and $p_3(x) = 2 + x$ form a linearly independent set in $V_3(x)$.

The set of polynomials with degree ≤ 2 forms a vector space. A set of vectors $\left\{ \vec{v}_1, \vec{v}_2, \ldots, \vec{v}_n \right\}$ is linearly independent if:

$$a_1\,\vec{v}_1 + a_2\,\vec{v}_2 + \ldots + a_n\,\vec{v}_n = 0$$

implies $a_1 = a_2 = \ldots = a_n = 0$ where $a_1, a_2, \ldots, a_n$ are scalars. Thus, let:

$$a_1 p_1(x) + a_2 p_2(x) + a_3 p_3(x) = 0 + 0x + 0x^2$$
$$a_1(x + 2x^2) + a_2(1 + 2x + x^2) + a_3(2 + x)$$
$$= 0 + 0x + 0x^2$$
$$a_1 x + 2a_1 x^2 + a_2 + 2a_2 x + a_2 x^2 + 2a_3 + a_3 x$$
$$= 0 + 0x + 0x^2$$

Gathering like terms,

$$(a_2 + 2a_3) + (a_1 + 2a_2 + 2a_3)x + (2a_1 + a_2)x^2$$
$$= 0 + 0x + 0x^2$$

Two polynomials are equal when the coefficients of like powers of x are equal. Thus,

$$(a_2 + 2a_3) = 0 \tag{1}$$
$$(a_1 + 2a_2 + a_3) = 0 \tag{2}$$
$$(2a_1 + a_2) = 0 \tag{3}$$

From (3) we get $a_2 = -2a_1$.

Substituting in (2), $a_1 - 4a_1 + a_3 = 0$ or $a_3 = 3a_1$.

Now substituting for a_2 and a_3 in (1), the result is $-2a_1 + 2(3a_1) = 0$. So, $4a_1 = 0$ implies $a_1 = 0$. Then $c_3 = 0$ and $a_2 = 0$.

We conclude that the polynomials are independent.

 Determine whether or not u and v are linearly dependent if:

a) $\vec{u} = (3, 4), \ \vec{v} = (1, -3)$

b) $\vec{u} = (2, -3), \ \vec{v} = (6, -9)$

c) $\vec{u} = (4, 3, -2), \ \vec{v} = (2, -6, 7)$

d) $\vec{u} = (-4, 6, -2), \ \vec{v} = (2, -3, 1)$

e) $\quad \vec{u} = \begin{pmatrix} 1 & -2 & 4 \\ 3 & 0 & -1 \end{pmatrix}, \quad \vec{v} = \begin{pmatrix} 2 & -4 & 8 \\ 6 & 0 & -2 \end{pmatrix}$

f) $\quad \vec{u} = \begin{pmatrix} 1 & 2 & -3 \\ 6 & -5 & 4 \end{pmatrix}, \quad \vec{v} = \begin{pmatrix} 6 & -5 & 4 \\ 1 & 2 & -3 \end{pmatrix}$

g) $\quad \vec{u} = 2 - 5t + 6t^2 - t^3, \ \vec{v} = 3 + 2t - 4t^2 + 5t^3$

h) $\quad \vec{u} = 1 - 3t + 2t^2 - 3t^3, \ \vec{v} = -3 + 9t - 6t^2 + 9t^3$

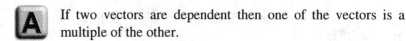 If two vectors are dependent then one of the vectors is a multiple of the other.

a) If $(3, 4)$ and $(1, -3)$ are dependent, then there exists a scalar a_1 such that $(3, 4) = a_1 (1, -3) = (a_1, -3a_1)$. (2)

In other words,

$$a_1 = 3$$
$$-3a_1 = 4$$

These two equations are inconsistent, i.e., there is no value of a_1 such that (2) holds.

Thus, $\vec{u}$ and $\vec{v}$ are linearly independent.

b) Using the same reasoning as in a), $\vec{u}$ and $\vec{v}$ are dependent if there exists an a_1 such that $a_1 \vec{u} = \vec{v}$ so,

$$a_1 (2, -3) = (6, -9)$$
$$2a_1 = 6 \qquad\qquad (2)$$
$$-3a_1 = -9 \qquad\qquad (3)$$

From (2), $a_1 = 3$ which, when substituted into (3), is true. Thus, $\vec{v} = 3\vec{u}$ and the two vectors are linearly dependent.

c)   We wish to find a_1 such that:

$$a_1 (4, 3, -2) = (2, -6, 7) \tag{4}$$

is true. Setting the corresponding coordinates of the triple of real numbers equal to each other, we obtain:

$$4a_1 = 2$$
$$3a_1 = -6 \tag{4'}$$
$$-2a_1 = 7$$

From the first equation of (4'), $a_1 = \frac{1}{2}$. From the second equation, $a_1 = -2$ and from the third, $a_1 = -\frac{7}{2}$. a_1 is clearly overworked. We conclude that the two vectors are linearly independent since (4) leads to a contradiction.

d)   Setting $a_1 (-4, 6, -2) = (2, -3, 1)$,

we have $-4a_1 = 2$

$$6a_1 = -3$$

$$-2a_1 = 1$$

The value $a_1 = -\frac{1}{2}$ satisfies all three equations. Hence, $\overrightarrow{v} = -\frac{1}{2}\overrightarrow{u}$ and the two vectors are linearly dependent.

e)   Here the vectors are 2×3 matrices, but vectors nevertheless. Hence, setting

$$a_1 \begin{pmatrix} 1 & -2 & 4 \\ 3 & 0 & -1 \end{pmatrix} = \begin{pmatrix} 2 & -4 & 8 \\ 6 & 0 & -2 \end{pmatrix},$$

$$\begin{pmatrix} a_1 & -2a_1 & 4a_1 \\ 3a_1 & 0 & -a_1 \end{pmatrix} = \begin{pmatrix} 2 & -4 & 8 \\ 6 & 0 & -2 \end{pmatrix}. \tag{5}$$

Two matrices $\{a_{ij}, b_{ij}\}$ are equal if and only if, $a_{ij} = b_{ij}$

$$(i = 1,\dots, m; j = 1,\dots, n).$$

Hence, (5) holds if and only if

$$a_1 = 2 \qquad\qquad 3a_1 = 6$$
$$-2a_1 = -4 \qquad\qquad 0 = 0$$
$$4a_1 = 8 \qquad\qquad -a_1 = -2$$

Since all of the above equalities are consistent, $a_1 = 2$ and $\overrightarrow{v} = 2\overrightarrow{u}$, thus, we have $\overrightarrow{v}$ is dependent on $\overrightarrow{u}$ (and vice-versa).

f) Suppose $\overrightarrow{u} = \begin{pmatrix} 1 & 2 & -3 \\ 6 & -5 & 4 \end{pmatrix}$ and $\overrightarrow{v} = \begin{pmatrix} 6 & -5 & 4 \\ 1 & 2 & -3 \end{pmatrix}$.

Then $a\overrightarrow{u} = \overrightarrow{v}$ implies

$$\begin{pmatrix} a & 2a & -3a \\ 6a & -5a & 4a \end{pmatrix} = \begin{pmatrix} 6 & -5 & 4 \\ 1 & 2 & -3 \end{pmatrix}$$

and gives $a = 6$ (1)

$2a = -5$, etc.   (2)

We do not need to go any further since we already have a contradiction, (2) gives $a = -5/2$ but (1) says $a = 6$. Since there exists no such a, $\overrightarrow{v}$ and $\overrightarrow{u}$ are linearly independent.

g) The set of all polynomials with degree $\leq n$ is a vector space. Here $\overrightarrow{u}$ and $\overrightarrow{v}$ are vectors from V_4. Setting:

$$a_1 (2 - 5t + 6t^2 - t^3) = (3 + 2t - 4t^2 + 5t^3),$$
$$2a_1 - 5a_1 t + 6a_1 t^2 - a_1 t^3 = 3 + 2t - 4t^2 + 5t^3.$$

Two polynomials are equal when coefficients of like powers of the variable are equal. Thus,

$$2a_1 = 3$$
$$-5a_1 = 2$$
$$6a_1 = -4$$
$$-a_1 = 5$$

Since a_1 cannot satisfy all these equations simultaneously, we conclude that it does not exist. Hence, $\overrightarrow{u}$ and $\overrightarrow{v}$ are not multiples of each other, i.e., they are independent.

h) Again, $\overrightarrow{u}$ and $\overrightarrow{v}$ are two vectors in a four-dimensional polynomial vector space. Let:

$$a_1(1 - 3t + 2t^2 - 3t^3) = -3 + 9t - 6t^2 + 9t^3$$
$$a_1 - 3a_1t + 2a_1t^2 - 3a_1t^3 = -3 + 9t - 6t^2 + 9t^3$$

Setting the coefficients of like powers of x equal to each other,

$$a_1 = -3$$
$$-3a_1 = 9$$
$$2a_1 = -6$$
$$-3a_1 = 9$$

$a_1 = -3$ satisfies all of the above equations. Hence, $\overrightarrow{v} = -3\,\overrightarrow{u}$ and $\overrightarrow{u}$ and $\overrightarrow{v}$ are dependent.

Write the vector $\overrightarrow{v} = (1, -2, 5)$ as a linear combination of the vectors $\overrightarrow{e}_1 = (1, 1, 1)$, $\overrightarrow{e}_2 = (1, 2, 3)$, and $\overrightarrow{e}_3 = (2, -1, 1)$.

We first show that the vectors e_1, e_2, and e_3 are linearly independent. Then we find the required constants.

A set of vectors $\left\{ \overrightarrow{e}_1, \overrightarrow{e}_2, \ldots, \overrightarrow{e}_n \right\}$ in R^n is linearly independent if, for $a_1 \in R$,

$$a_1 \overrightarrow{e}_1 + a_2 \overrightarrow{e}_2 + \ldots + a_n \overrightarrow{e}_n = 0$$

implies $a_1 = 0$ ($i = 1, \ldots, n$). Setting:

$$a_1 \overrightarrow{e}_1 + a_2 \overrightarrow{e}_2 + a_3 \overrightarrow{e}_3 = (0, 0, 0),$$
$$a_1 (1, 1, 1) + a_2 (1, 2, 3) + a_3 (2, -1, 1) = (0, 0, 0)$$
$$(a_1, a_1, a_1) + (a_2, 2a_2, 3a_2) + (2a_3, -a_3, a_3) = (0, 0, 0).$$

Adding coordinates,

$$(a_1 + a_2 + 2a_3, a_1 + 2a_2 - a_3, a_1 + 3a_2 + a_3) = (0, 0, 0).$$

Two points in R^3 are equal only if their respective coordinates are equal. Hence,

$$L_1: a_1 + a_2 + 2a_3 = 0 \tag{1}$$
$$L_2: a_1 + 2a_2 - a_3 = 0$$
$$L_3: a_1 + 3a_2 + a_3 = 0$$

We can reduce this system to row-echelon form as follows:

Replace L_2 by $L_1 - L_2$ and replace L_3 by $L_1 - L_3$.

$$L_1: a_1 + a_2 + 2a_3 = 0$$
$$L_2: 0 - a_2 - 3a_3 = 0$$
$$L_3: 0 - 2a_2 + a_3 = 0$$

Now replace L_3 by (-2) times L_2 added to L_3. We arrive at

$$a_1 + a_2 + 2a_3 = 0$$
$$0 - a_2 + 3a_3 = 0$$
$$0 + 0 - 5a_3 = 0$$

Thus, $a_1 = a_2 = a_3 = 0$ is the only solution and $\left\{ \vec{e_1}, \vec{e_2}, ..., \vec{e_3} \right\}$ is linearly independent. This means that the given $\vec{v}$ can be written as a linear combination of $\vec{e}_1$, $\vec{e}_2$, and $\vec{e}_3$.

Let $\vec{v} = a_1 \vec{e}_1 + a_2 \vec{e}_2 + a_3 \vec{e}_3$.

$(1, -2, 5) = a_1 (1, 1, 1) + a_2 (1, 2, 3) + a_3 (2, -1, 1)$

$(1, -2, 5) = (a_1, a_1, a_1) + (a_2, 2a_2 + 3a_2) + (2a_3, -a_3, a_3)$

$(1, -2, 5) = (a_1 + a_2 + 2a_3, a_1 + 2a_2 - a_3, a_1 + 3a_2 + a_3)$

Setting corresponding coordinates equal to each other,

$$L_1: a_1 + a_2 + 2a_3 = 1 \tag{2}$$
$$L_2: a_1 + 2a_2 - a_3 = -2$$
$$L_3: a_1 + 3a_2 + a_3 = 5$$

We can apply row operations to the non-homogeneous system (2) to reduce it to row-echelon form. System (2) becomes:

$$L_1: a_1 + a_2 + 2a_3 = 1$$
$$L_2: 0 + a_2 - 3a_3 = -3$$
$$L_3: 0 + 2a_2 - a_3 = 4$$

Multiply L_1 by -1 and add it to L_2. Also, multiply L_1 by -1 and add it to L_3.

$$a_1 + a_2 + 2a_3 = 1$$
$$0 + a_2 - 3a_3 = -3$$
$$0 + 0 + 5a_3 = 10$$

Multiply L_2 by -2 and add to L_3, the results of which are given above.

Solving by back-substitution,

$$a_3 = 2;\ a_2 = 3;\ a_1 = -6.$$

Hence, $\vec{v} = -6\,\vec{e}_1 + 3\,\vec{e}_2 + 2\,\vec{e}_3.$

 Let the functions 1, x, and x^2 be defined on the interval $[0, 1]$. Then these functions belong to the vector space of continuous, real-valued functions defined on $[0, 1]$ called $C[0, 1]$. Show that 1, x, and x^2 are independent.

A set of vectors W = $\left\{ \vec{w}_1, \ldots, \vec{w}_n \right\}$ is linearly independent if:

$$a_1\,\vec{w}_1 + a_2\,\vec{w}_2 + \ldots + a_n\,\vec{w}_n = 0$$

implies $a_i = 0$ $(i = 1, \ldots, n)$.

Thus, we set

$$a_1(1) + a_2(x) + a_3(x^2) = 0 \qquad (1)$$

If (1) implies $a_1 = a_2 = a_3 = 0$ for all $0 \le x \le 1$, then we can conclude that 1, x, and x^2 are linearly independent. Suppose $x = 0$ in (1). Then, $c_1 = 0$. Now, differentiate (1):

$$a_2 + 2a_3 x = 0 \tag{2}$$

Setting $x = 0$ in (2) yields $a_2 = 0$. Similarly, differentiating (2),

$$2a_3 = 0.$$

We conclude that the only solution to (1) is $a_1 = a_2 = a_3 = 0$, and, hence, the functions 1, x, and x^2 are independent.

 Determine whether the vectors $\vec{x}$, $\vec{y}$ and $\vec{z}$ are dependent or independent where:

a) $\vec{x} = (1, 1, -1)$, $\vec{y} = (2, -3, 1)$ $\vec{z} = (8, -7, 1)$

b) $\vec{x} = (1, -2, -3)$, $\vec{y} = (2, 3, -1)$, $\vec{z} = (3, 2, 1)$

c) $\vec{x} = (x_1, x_2)$, $\vec{y} = (y_1, y_2)$, $\vec{z} = (z_1, z_2)$

 A set of vectors $\left\{ \vec{v_1}, \vec{v_2}, \ldots \vec{v_n} \right\}$ is said to be linearly dependent if there exists a_i $(i = 1, 2, \ldots, n)$ such that:

$$a_1 \vec{v_1} + a_2 \vec{v_2} + \ldots + a_n \vec{v_n} = 0. \tag{1}$$

a)   Let $a_1 \vec{x} + a_2 \vec{y} + a_3 \vec{z} = 0$;

$a_1 (1, 1, -1) + a_2 (2, -3, 1) + a_3 (8, -7, 1) = (0, 0, 0)$,
$(a_1, a_1, -a_1) + (2a_2, -3a_2, a_2) + (8a_3, -7a_3, a_3) =$
$(0, 0, 0)$

or

$$a_1 + 2a_2 + 8a_3 = 0 \tag{2}$$
$$a_1 - 3a_2 - 7a_3 = 0$$
$$-a_1 + a_2 + a_3 = 0$$

The system (2) may now be reduced to echelon form. Constructing the coefficient matrix from (2), where T_i are the row operations as follows:

T_1: Replace the third row with the sum of the first and the third rows. Also, replace the second row with the second row minus the first row.

T_2: Replace the second row with $(^{-1}\!/_5)$ times the second row and replace the third row with $(^1\!/_3)$ times the third row.

T_3: The third row is the same as the second and so it may be dropped.

$$\begin{pmatrix} 1 & 2 & 8 \\ 1 & -3 & -7 \\ -1 & 1 & 1 \end{pmatrix} \xleftarrow{\;T_1\;} \begin{pmatrix} 1 & 2 & 8 \\ 0 & -5 & -15 \\ 0 & 3 & 9 \end{pmatrix} \xleftarrow{\;T_2\;} \begin{pmatrix} 1 & 2 & 8 \\ 0 & 1 & 3 \\ 0 & 1 & 3 \end{pmatrix} \xleftarrow{\;T_3\;} \begin{pmatrix} 1 & 2 & 8 \\ 0 & 1 & 3 \end{pmatrix}.$$

Thus,

$$a_1 + 2a_2 + 8a_3 = 0 \tag{3}$$
$$a_2 + 3a_3 = 0$$

The system (3), equivalent to system (2), has two equations in three unknowns. The second equation in (3) yields $a_2 = -3a_3$. Substituting in the first equation, we get

$$a_1 + 2(-3a_3) + 8a_3 = 0,$$
so
$$a_1 = -2a_3.$$

Thus, all of the solutions of this system are:

$$a_1 = -2a$$
$$a_2 = -3a$$

where:

$$a_3 = a$$

a is any scalar. If $a = 1$, one non-zero solution is $a_1 = -2$, $a_2 = -3$, $a_3 = 1$. Hence, the vectors are dependent.

b) Let $a_1 \overrightarrow{x} + a_2 \overrightarrow{y} + a_3 \overrightarrow{z} = 0$;

$$(a_1, -2a_1 - 3a_1) + (2a_2, 3a_2 - a_2) + (3a_3, 2a_3, a_3) = (0, 0, 0)$$

or

$$a_1 + 2a_2 + 3a_3 = 0 \tag{4}$$
$$-2a_1 + 3a_2 + 2a_3 = 0$$
$$-3a_1 - a_2 + a_3 = 0$$

Reducing the coefficient matrix of (4) to echelon form using appropriate row operations:

T_1: Replace the second row with two times the first row plus the second row; replace the third row with three times the first row plus the third row.

T_2: Replace the third row with the sum of -5 times the second row and 7 times the third row.

$$\begin{pmatrix} 1 & 2 & 3 \\ -2 & 3 & 2 \\ -3 & -1 & 1 \end{pmatrix} \xleftarrow{T_1} \begin{pmatrix} 1 & 2 & 3 \\ 0 & 7 & 8 \\ 0 & 5 & 10 \end{pmatrix} \xleftarrow{T_2} \begin{pmatrix} 1 & 2 & 3 \\ 0 & 7 & 8 \\ 0 & 0 & 30 \end{pmatrix}$$

where T_i are elementary row operations.

The system $a_1 + 2a_2 + 3a_3 = 0$
$$7a_2 + 8a_3 = 0$$
$$30a_3 = 0$$

has only the trivial solution $a_1 = a_2 = a_3 = 0$. Thus, by definition, the vectors are linearly independent.

c) Note here that $\vec{x}, \vec{y}, \vec{z}$ are arbitrary vectors in R^2.

Then let $a_1(x_1, x_2) + a_2(y_1, y_2) + a_3(z_1, z_2) = (0, 0)$.

$$\begin{pmatrix} x_1 & y_1 & z_1 \\ x_2 & y_2 & z_2 \end{pmatrix} \begin{pmatrix} a_1 \\ a_2 \\ a_3 \end{pmatrix} = \begin{pmatrix} 0 \\ 0 \end{pmatrix}$$

or

$$x_1 a_1 + y_1 a_2 + z_1 a_3 = 0 \tag{5}$$
$$x_2 a_1 + y_2 a_2 + z_2 a_3 = 0$$

Because there are more unknowns (a_1, a_2, a_3) than equations, the system has a non-zero solution.

Since $\vec{x}$, $\vec{y}$, $\vec{z}$ were arbitrary vectors, this shows that any three vectors in R^2 are linearly dependent. In general, any $n + 1$ vectors in R^n are linearly dependent.

3.5 Basis and Dimension

If V is any vector space and $S = \left\{ \vec{v_1}, \vec{v_2}, \vec{v_3}, \ldots, \vec{v_r} \right\}$ is a finite set of vectors in V, then s is called a basis for V. If:

a) S is linearly independent

b) S spans V

EXAMPLE

$v = \left\{ (1,0,0), (0,1,0), (0,0,1) \right\}$ is the basis for R^3.

The set of vectors $\left\{ (1,0,\ldots,0), (0,1,0,\ldots,0), \ldots, (0,0,\ldots,1) \right\}$ in R^n is called the standard basis for R^n.

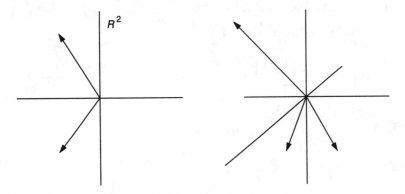

Figure 3.7 Any two non-collinear vectors in R^2 span R^2.

Figure 3.8 Any three non-coplanar vectors in R^3 span R^3.

If a non-zero vector space contains a basis consisting of a finite set of vectors, then it is called finite dimensional. Otherwise, it is infinite dimensional. The zero vector space is defined to be finite dimensional.

THEOREM

Any two bases for a finite dimensional vector space must have the same number of vectors.

The number of vectors in a basis of a finite dimensional vector space is called the dimension of that vector space. The zero vector space is said to have dimension zero.

THEOREM

If $S = \{v_1, v_2,..., v_r\}$ is a linearly independent set of vectors in an n-dimensional space V, and $r < n$, then S can be enlarged to form a basis for V.

Problem Solving Examples:

 Find the dimension of the vector space spanned by:

a) $(1, -2, 3, -1)$ and $(1, 1, -2, 3)$

b) $(3, -6, 3, -9)$ and $(-2, 4, -2, 6)$

c) $t^3 + 2t^2 + 3t + 1$ and $2t^3 + 4t^2 + 6t + 2$

d) $t^3 - 2t^2 + 5$ and $t^2 + 3t - 4$

e) $\begin{bmatrix} 1 & 2 \\ 1 & 2 \end{bmatrix}$ and $\begin{bmatrix} 1 & 1 \\ 2 & 2 \end{bmatrix}$

f) $\begin{bmatrix} 1 & 1 \\ -1 & -1 \end{bmatrix}$ $\begin{bmatrix} -3 & -3 \\ 3 & 3 \end{bmatrix}$

g) 3 and -3

A   Two non-zero vectors span a space W of dimension two if they are independent, and of dimension one if they are dependent. Two vectors are dependent if and only if one is a scalar multiple of the other. Now, using the above facts, the dimension of the given vector space can be found. Hence, the dimensions of the subspaces spanned by the given sets of vectors are, respectively,

a) 2, b) 1, c) 1, d) 2, e) 2, f) 1, and g) 1.

Note that in a) and b) the subspace spanned is a subspace of the vector space R^4; in c) and d) it is a subspace of the real vector space of polynomials in t over the field R; in e) and f) it is a subspace of the real vector space of 2×2 matrices with entries in R; and in g) it is just R considered as a real vector space.

 Show that (1, 2, 1), (1, –1, 3), and (1, 1, 4) form a basis for R^3.

A basis for a vector space is the smallest set of vectors from the space that spans the space. In the given problem this means that any vector in R^3 (x, y, z) must be uniquely expressible as a linear combination of (1, 2, 1), (1, –1, 3), and (1, 1, 4). Thus, we must show that for any (x, y, z) there exists a_1, a_2, a_3 such that:

$$\begin{bmatrix} x \\ y \\ z \end{bmatrix} = a_1 \begin{bmatrix} 1 \\ 2 \\ 1 \end{bmatrix} + a_2 \begin{bmatrix} 1 \\ -1 \\ 3 \end{bmatrix} + a_3 \begin{bmatrix} 1 \\ 1 \\ 4 \end{bmatrix}. \tag{1}$$

Expanding (1), we obtain the three simultaneous equations:

$$\begin{aligned} x &= a_1 + a_2 + a_3 \\ y &= 2a_1 - a_2 + a_3 \\ z &= a_1 + 3a_2 + 4a_3 \end{aligned} \tag{2}$$

System (2) can be rewritten in matrix form as:

$$\begin{bmatrix} 1 & 1 & 1 \\ 2 & -1 & 1 \\ 1 & 3 & 4 \end{bmatrix} \begin{bmatrix} a_1 \\ a_2 \\ a_3 \end{bmatrix} = \begin{bmatrix} x \\ y \\ z \end{bmatrix}. \tag{3}$$

If the coefficient matrix of the a-vector is invertible, its inverse premultiplied by $[x\ y\ z]^t$ will yield the unique solution for $[a_1\ a_2\ a_3]^t$. To find the required inverse if it exists, we form the augmented matrix:

$$\begin{bmatrix} 1 & 1 & 1 & 1 & 0 & 0 \\ 2 & -1 & 1 & 0 & 1 & 0 \\ 1 & 3 & 4 & 0 & 0 & 1 \end{bmatrix}. \tag{4}$$

We use the elementary row operations to change the left part of (4) to the identity matrix. Simultaneously, the identity matrix on the right will change to the required inverse. Symbolically, $[A:I] \leftrightarrow [I:A^{-1}]$.

Replace row 3 with row 3 – row 1.

Replace row 2 with $2 \times$ row 1 – row 2.

$$\begin{bmatrix} 1 & 1 & 1 & 1 & 0 & 0 \\ 0 & 3 & 1 & 2 & -1 & 0 \\ 0 & 2 & 3 & -1 & 0 & 1 \end{bmatrix}$$

Replace row 2 by $\frac{1}{3} \times$ row 2.

$$\begin{bmatrix} 1 & 1 & 1 & 1 & 0 & 0 \\ 0 & 1 & \frac{1}{3} & \frac{2}{3} & -\frac{1}{3} & 0 \\ 0 & 2 & 3 & -1 & 0 & 1 \end{bmatrix}$$

Replace row 1 with row 1 – row 2.

Replace row 3 with row 3 – $2 \times$ row 2.

$$\begin{bmatrix} 1 & 0 & \frac{2}{3} & \frac{1}{3} & \frac{1}{3} & 0 \\ 0 & 1 & \frac{1}{3} & \frac{2}{3} & -\frac{1}{3} & 0 \\ 0 & 0 & \frac{7}{3} & -\frac{7}{3} & \frac{2}{3} & 1 \end{bmatrix}$$

Replace row 3 by $\frac{3}{7} \times$ row 3.

$$\begin{bmatrix} 1 & 0 & \frac{2}{3} & \frac{1}{3} & \frac{1}{3} & 0 \\ 0 & 1 & \frac{1}{3} & \frac{2}{3} & -\frac{1}{3} & 0 \\ 0 & 0 & 1 & -1 & \frac{2}{7} & \frac{3}{7} \end{bmatrix}$$

Replace row 1 by row 1 – $\frac{2}{3}$ row 3.

Replace row 2 by row $2 - \frac{1}{3}$ row 3.

And, therefore, the matrix (4) becomes:

$$\begin{bmatrix} 1 & 0 & 0 \\ 0 & 1 & 0 \\ 0 & 0 & 1 \end{bmatrix} \left| \begin{array}{ccc} 1 & \frac{1}{7} & -\frac{2}{7} \\ 1 & -\frac{3}{7} & -\frac{1}{7} \\ -1 & \frac{2}{7} & \frac{3}{7} \end{array} \right. . \tag{5}$$

Premultiplying (3) by the inverse given by (5),

$$a_1 = x + \frac{1}{7}y - \frac{2}{7}z \tag{6}$$

$$a_2 = x - \frac{3}{7}y - \frac{1}{7}z$$

$$a_3 = -x + \frac{2}{7}y + \frac{3}{7}z$$

Any vector in R^3, $[x, y, z]$ can be expressed in the form (1) with coefficients given by (6). To clarify the result, let $[x\ y\ z]' = [1\ 2\ 3]'$.

Then

$$a_1 = 1 + \frac{1}{7}(2) - \frac{2}{7}(3) = \frac{3}{7}$$

$$a_2 = 1 - \frac{3}{7}(2) - \frac{1}{7}(3) = -\frac{2}{7}$$

$$a_3 = -1 + \frac{2}{7}(2) + \frac{3}{7}(3) = \frac{6}{7}$$

and $\dfrac{3}{7}\begin{bmatrix} 1 \\ 2 \\ 1 \end{bmatrix} - \dfrac{2}{7}\begin{bmatrix} 1 \\ -1 \\ 3 \end{bmatrix} + \dfrac{6}{7}\begin{bmatrix} 1 \\ 1 \\ 4 \end{bmatrix} = \begin{bmatrix} 1 \\ 2 \\ 3 \end{bmatrix} = [x\ y\ z]'.$

Thus, $[1\ 2\ 1]'$, $[1\ -1\ 3]'$, and $[1\ 1\ 4]'$ form a basis for R^3 with coordinates given by (6).

Q What is the usual basis for the vector space V of all real 2×3 matrices?

A A basis B for a vector space V is a set of vectors in V such that any element in V can be expressed as a linear combination of vectors in B. Further, B is the smallest such set. (By requiring it to be the smallest such set, we secure the linear independence of its elements.)

An arbitrary, real 2×3 matrix has the form:

$$A = \begin{bmatrix} a_{11} & a_{12} & a_{13} \\ a_{21} & a_{22} & a_{23} \end{bmatrix}. \quad a_{ij} \in \text{Real numbers} \qquad (1)$$

Since each element of the matrix is arbitrary, we need at least six matrices to express A as a linear combination. The most natural choice of matrices is:

$$A_1 = \begin{bmatrix} 1 & 0 & 0 \\ 0 & 0 & 0 \end{bmatrix}; \quad A_2 = \begin{bmatrix} 0 & 1 & 0 \\ 0 & 0 & 0 \end{bmatrix}; \dots, \quad A_6 = \begin{bmatrix} 0 & 0 & 0 \\ 0 & 0 & 1 \end{bmatrix}.$$

Then $A = a_{11} A_1 + a_{12} A_2 + a_{13} A_3 + \dots + a_{23} A_6$. The set of matrices $\{A_1, \dots, A_6\}$ is called the usual basis for the vector space of 2×3 matrices.

For example, let $A = \begin{bmatrix} 14 & -7 & 6 \\ 3 & 0 & 5 \end{bmatrix}$.

Then $A = 14A_1 - 7A_2 + 6A_3 + 3A_4 + 0A_5 + 5A_6$.

The dimension of a vector space is the number of elements in the basis. We see that $V_{2 \times 3}$ has dimension six. In general, $V_{m \times n}$ has dimension mn.

Q One basis for the vector space V of polynomials of degree not exceeding two is the set $B = \{1, x, x^2\}$. Construct another basis for V whose elements are all quadratic functions.

Since B contains three vectors, we know that any other set containing three linearly independent vectors will be a basis. Consider the three vectors:

$$f_1 = \frac{(x-1)(x-2)}{2}; \ f_2 = \frac{x(x-2)}{-1}; \ f_3 = \frac{x(x-1)}{2}. \quad (1)$$

Note that:

$$\begin{aligned} f_1(1) = f_1(2) = 0; \ f_1(0) = 1 \\ f_2(0) = f_2(2) = 0; \ f_2(1) = 1 \\ f_3(0) = f_3(1) = 0; \ f_3(2) = 1 \end{aligned} \quad (2)$$

The polynomials f_1, f_2, and f_3 each have degree 2 and thus belong to V.

Let us check for linear independence. Set:

$$a_1 f_1(x) + a_2 f_2(x) + a_3 f_3(x) = 0. \quad (3)$$

Substitute $x = 0$ in (3); from the relations (2),

$$a_1(1) + a_2(0) + a_3(0) = 0$$

implies $a_1 = 0$.

Now, let $x = 1$ in (3). Then, using (2),

$$0 + a_2(1) + a_3(0) = 0$$

implies $a_2 = 0$. Similarly, letting $x = 2$ in (3) and using (2), we find that $a_3 = 0$. So (3) implies that $a_1 = a_2 = a_3 = 0$. Thus, $f_1(x), f_2(x)$, and $f_3(x)$ are linearly independent and form a basis for V.

To find the coordinates of an arbitrary vector in V, let:

$$f(x) = a_1 f_1(x) + a_2 f_2(x) + a_3 f_3(x). \quad (4)$$

We can use the relations (2) one more time. When $x = 0$, $a_1 = f(0)$. For $x = 1$, $a_2 = f(1)$, and when $x = 2$, $a_3 = f(2)$. Thus,

$$f(x) = f(0)f_1(x) + f(1)f_2(x) + f(2)f_3(x). \quad (5)$$

The formula (5) is known as the Lagrange interpolation formula.

3.6 Row and Column Space of a Matrix

Given the $m \times n$ matrix

$$A = \begin{bmatrix} a_{11} & a_{12} & \cdots & a_{1n} \\ a_{21} & a_{22} & \cdots & a_{2n} \\ \vdots & \vdots & & \vdots \\ a_{m1} & a_{m2} & \cdots & a_{mn} \end{bmatrix}$$

a) The vectors $(a_{11}, a_{12}, \ldots, a_{1n}), (a_{21}, a_{22}, \ldots, a_{2n}), \ldots, (a_{m1}, a_{m2}, \ldots, a_{mn})$ are called the row vectors of A.

b) The vectors $(a_{11}, a_{21}, \ldots, a_{m1})$, $(a_{12}, a_{22}, \ldots, a_{m2}), \ldots,$ $(a_{1n}, a_{2n}, \ldots, a_{mn})$ are called the column vectors of A.

c) The row space of A is the subspace of R^n spanned by the row vectors.

d) The column space of A is the subspace of R^m spanned by the column vectors.

THEOREM

Elementary row operations do not change the row space of a matrix.

THEOREM

The non-zero row vectors from the row-echelon form of a matrix form a basis for the row space of that matrix.

THEOREM

The row space and column space of a matrix have the same dimension.

The rank of a matrix is defined to be the dimension of the row space and column space of that matrix.

EXAMPLE

The rank of $\begin{bmatrix} 1 & 0 & 1 & 1 \\ 3 & 2 & 5 & 1 \\ 0 & 4 & 4 & -4 \end{bmatrix}$ is 2.

THEOREM

If A is an $n \times n$ matrix, then the following statements are equivalent:

a) A is invertible.

b) $A\vec{X} = \vec{0}$ has the only trivial solution.

c) A is row-equivalent to the $n \times n$ identity matrix.

d) $A\vec{X} = \vec{b}$ is consistent for every $n \times 1$ matrix b.

e) $\det(A) \neq 0$.

f) A has rank n.

g) The row and column vectors of A are linearly independent.

Problem Solving Examples:

 Find the rank of the matrix A where:

a) $$A = \begin{bmatrix} 1 & 3 & 1 & -2 & -3 \\ 1 & 4 & 3 & -1 & -4 \\ 2 & 3 & -4 & -7 & -3 \\ 3 & 8 & 1 & -7 & -8 \end{bmatrix}$$

b) $$A = \begin{bmatrix} 1 & 2 & -3 \\ 2 & 1 & 0 \\ -2 & -1 & 3 \\ -1 & 4 & -2 \end{bmatrix}$$

c) $$A = \begin{bmatrix} 1 & 3 \\ 0 & -2 \\ 5 & -1 \\ -2 & 3 \end{bmatrix}$$

(a) First, reduce the matrix A to echelon form using the elementary row operations.

(b)   Add -1 times the first row to the second row.

(c)   Add -2 times the first row to the third row.

(d)   Add -3 times the first row to the third row.

$$A = \begin{bmatrix} 1 & 3 & 1 & -2 & -3 \\ 0 & 1 & 2 & 1 & -1 \\ 0 & -3 & -6 & -3 & 3 \\ 0 & -1 & -2 & -1 & 1 \end{bmatrix}$$

Add $+3$ times the second row to the third row.

Add the second row to the fourth row. Then,

$$A = \begin{bmatrix} 1 & 3 & 1 & -2 & -3 \\ 0 & 1 & 2 & 1 & -1 \\ 0 & 0 & 0 & 0 & 0 \\ 0 & 0 & 0 & 0 & 0 \end{bmatrix}.$$

Since the echelon matrix has two nonzero rows, rank $(A) = 2$.

(b)   Since row rank equals column rank, it is easier to form the transpose of A and then row reduce to echelon form.

$$A = \begin{bmatrix} 1 & 2 & -2 & -1 \\ 2 & 1 & -1 & 4 \\ -3 & 0 & 3 & -2 \end{bmatrix}$$

Add -2 times the first row to the second row, and add 3 times the first row to the third row.

$$A = \begin{bmatrix} 1 & 2 & -2 & -1 \\ 0 & -3 & 3 & 6 \\ 0 & 6 & -3 & -5 \end{bmatrix}$$

Add 2 times the second row to the third row.

$$A = \begin{bmatrix} 1 & 2 & -2 & -1 \\ 0 & -3 & 3 & 6 \\ 0 & 0 & 3 & 7 \end{bmatrix}$$

Since the echelon matrix has three non-zero rows, rank $(A) = 3$.

c)   The two columns are linearly independent since one is not a multiple of the other. Hence, rank $(A) = 2$.

    Find the rank of the matrix A where

$$A = \begin{bmatrix} 1 & 0 & 2 & 3 \\ 0 & 0 & 5 & 1 \\ 0 & 0 & 0 & 0 \end{bmatrix}.$$

    The matrix A is an echelon matrix. The rank of an echelon matrix is the number of non-zero row vectors in the matrix. This follows from the fact that the non-zero rows of an echelon matrix are linearly independent.

Since the echelon matrix A has two non-zero row vectors, its rank is 2.

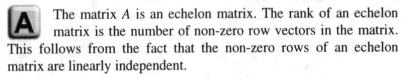   Consider the system of equations

$$\begin{bmatrix} 2 & 1 & 3 \\ 1 & -2 & 2 \\ 0 & 1 & 3 \end{bmatrix} \begin{bmatrix} x_1 \\ x_2 \\ x_3 \end{bmatrix} = \begin{bmatrix} 1 \\ 2 \\ 3 \end{bmatrix}.$$

Show that the system has a solution without actually computing a solution.

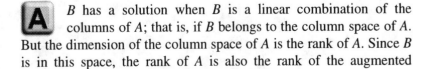 B has a solution when B is a linear combination of the columns of A; that is, if B belongs to the column space of A. But the dimension of the column space of A is the rank of A. Since B is in this space, the rank of A is also the rank of the augmented

matrix $[A \mid B]$. Thus, a necessary and sufficient condition for $AX = B$ to have a solution is that rank $(A) = $ rank $[A \mid B]$.

Applying elementary row operations to the given matrix, it can be seen that:

$$\begin{bmatrix} 2 & 1 & 3 \\ 1 & -2 & 2 \\ 0 & 1 & 3 \end{bmatrix} \tag{1}$$

is equivalent to:

$$\begin{bmatrix} 1 & 0 & 8 \\ 0 & 1 & 3 \\ 0 & 0 & -16 \end{bmatrix} \tag{2}$$

Since the columns of the matrix (2) form a basis for R^3, rank $(A) = 3$. Next, form the augmented matrix:

$$\left[\begin{array}{ccc|c} 2 & 1 & 3 & 1 \\ 1 & -2 & 2 & 2 \\ 0 & 1 & 3 & 3 \end{array}\right]$$

and reduce to echelon form:

$$\left[\begin{array}{ccc|c} 1 & 0 & 0 & -1 \\ 0 & 1 & 0 & -\frac{3}{8} \\ 0 & 0 & 1 & \frac{9}{8} \end{array}\right]$$

Thus, rank $(A \mid B) = 3$ and the given system of equations has a solution.

3.7 Inner Product Spaces

An inner product on a vector space V produces a real number $< \vec{v_1}, \vec{v_2} >$ from each pair of vectors $\vec{v_1}, \vec{v_2} \in V$, which satisfies the following axioms for all vectors $\vec{v_1}, \vec{v_2}, \vec{v_3} \in V$, and for all scalars L.

a) $< \vec{v_1}, \vec{v_2} > = < \vec{v_2}, \vec{v_1} >$

b) $< \vec{v_1}, \vec{v_2}, \vec{v_3}, > = < \vec{v_1}, \vec{v_3}, > + < \vec{v_2}, \vec{v_3}, >$

c) $< L\vec{v_1}, \vec{v_2} > = L < \vec{v_1}, \vec{v_2}, >$

d) $< \vec{v_1}, \vec{v_1} > \geq$ and $< \vec{v_1}, \vec{v_1}, > = 0$, if and only if $\vec{v} = \vec{0}$.

If $\vec{X} = (x_1, x_2, \ldots, x_n)$ and $\vec{Y} = (y_1, y_2, \ldots, y_n)$ are vectors in R^n, then the Euclidean inner product is defined as:

$$< \vec{X}, \vec{Y} > = X_1 Y_1 + X_2 Y_2 + \ldots + X_n Y_n$$

EXAMPLE

If $X = (2, 4, 6)$ and $Y = (7, 1, 1)$, then:

$<X, Y> = 14 + 4 + 6 = 24.$

A vector space with an inner product is called an inner product space.

3.7.1 Cauchy-Schwarz Inequality

If a and b are vectors in an inner product space, then:

$$< \vec{a}, \vec{b} >^2 \leq < \vec{a}, \vec{a} > < \vec{b}, \vec{b} >$$

Problem Solving Examples:

Compute $< \vec{u}, \vec{v} >$ where a) $\vec{u} = (2, -3, 6)$, $\vec{v} = (8, 2, -3)$; b) $\vec{u} = (1, -8, 0, 5)$, $\vec{v} = (3, 6, 4)$; c) $\vec{u} = (3, -5, 2, 1)$, $\vec{v} = (4, 1, -2, 5)$.

To compute the inner product of two vectors from R^n, multiply corresponding components and add.

a) $<\vec{u}, \vec{v}> = (2, -3, 6) \times (8, 2, -3) = 2(8) - 3(2) + 6(-3) = -8.$

b) $<\vec{u}, \vec{v}> = (1, -8, 0\ 5) \times (3, 6, 4).$ But, here the dot product is not defined since $u \in R^4$ while $v \in R^3$.

c) $<\vec{u}, \vec{v}> = (3, -5, 2, 1) \times (4, 1, -2, 5) = 3(4) - 5(1) + 2(-2) + 1(5) = 8.$

Let $(x_1, x_2, ..., x_n)$, $(y_1, y_2, ..., y_n)$, $(z_1, z_2, ..., z_n)$ be three vectors in R^n. Verify the properties of the inner product using these vectors.

a) $<X, Y> = <(x_1, x_2, ..., x_n), (y_1, y_2, ..., y_n)>$
$= x_1 y_1 + x_2 y_2 + ... + x_n y_n$
$= y_1 x_1 + y_2 x_2 + ... + y_n x_n$
$= <(y_1, y_2, ..., y_n), (x_1, x_2, ..., x_n)>$
$= <Y, X>$

b) $<(X + Y), Z> = <[(x_1, x_2, ..., x_n) + (y_1, y_2, ..., y_n)], (z_1, z_2, ..., z_n)>$
$= <(x_1 + y_1, x_2 + y_2, ..., x_n + y_n), (z_1, z_2, ..., z_n)>$
$= [(x_1 + y_1)z_1 + (x_2 + y_2)z_2 + ... + (x_n + y_n)z_n]$
$= [(x_1 z_1 + y_1 z_1) + (x_2 z_2 + y_2 z_2) + ... + (x_n z_n + y_n z_n)]$

(by the associative property of real numbers)

$= (x_1 z_1 + x_2 z_2 + ... + x_n z_n + y_1 z_1 + y_2 z_2 + ... + y_n z_n)$
$= (x_1 z_1 + x_2 z_2 + ... + x_n z_n) + (y_1 z_1 + y_2 z_2 + ... + y_n z_n)$
$= <X, Z> + <Y, Z>$

c) Here, L is a scalar from the field over which R^n is defined.

$(LX) \times Y = <[L(x_1, x_2, ..., x_n)], (y_1, y_2, ..., y_n)>$
$= <(Lx_1, Lx_2, ..., Lx_n), (y_1, y_2, ..., y_n)>$
$= Lx_1 y_1 + Lx_2 y_2 + ... + Lx_n y_n$

d) $<X, X> = <(x_1, x_2,\ldots, x_n), (x_1, x_2,\ldots, x_n)>$

$$= x_1^2 + x_2^2 + \ldots + x_n^2$$

$$= \sum_{i=1}^{n} x_i^2 \geq 0$$

Let P_n be the space of polynomials of degree less than n in a real variable. Define an inner product on P_n and find the inner product of the two functions $p(x) = x^2$ and $q(x) = 1 - x$.

An inner product for P_n is

$$< f(x), g(x) > = \int_0^1 f(x)g(x)dx. \qquad (1)$$

Each of the requirements for an inner product must be checked. First, realize that $f(x)$ and $g(x)$ are polynomials (therefore, integrable) which implies that the definite integral in (1) is a real number. Refer to a) – d) on page 113. Next, a) is satisfied since

$$< f(x), g(x) > = \int_0^1 f(x)g(x)dx = \int_0^1 g(x)f(x)dx = < g(x), f(x) > .$$

Now,

$$< f(x), g(x) + w(x) > = \int_0^1 f(x)[g(x) + w(x)]dx = \int_0^1 f(x)g(x) + f(x)w(x)dx$$

$$= \int_0^1 f(x)g(x)dx + \int_0^1 f(x)w(x)dx$$

$$= < f(x), g(x) > + < f(x), w(x) >$$

therefore, b) is satisfied. Next, $<f(x), Lg(x)>$ (L is a real number)

$$= \int_0^1 f(x) < g(x)dx = L > < f(x), g(x) > .$$

Lastly investigate d). For any polynomial function, $f(x)$, we know $(f(x))^2 \geq 0$. If $f(x)$ is not constantly 0 on the interval [0, 1],

$$\int_0^1 (f(x))^2 \, dx \neq 0.$$

To see this, recall that this integral is the area under the curve $[f(x)]^2$ from $x = 0$ to 1. So, $f(x)$, not always 0 on [0, 1], implies $<f(x), f(x)> > 0$. If $f(x) = 0$, $(f(x))^2 = 0$ and thus,

$$= \int_0^1 (f(x))^2\, dx = \int_0^1 0\, dx = 0.$$

Thus, (1) defines an inner product.

3.8 Length and Angle in Inner Product Spaces

If V is an inner product space, then the norm (or length) of a vector $\vec{a} \in V$ is defined as:

$$\left\| \vec{a} \right\| = <\vec{a}, \vec{a}>^{\frac{1}{2}}$$

EXAMPLE

If $\vec{a} = (3, 2, 6)$, then $\left\| \vec{a} \right\| = 7$.

If V is an inner product space, then the distance between two vectors $a, b, \in V$ is defined as:

$$d\left(\vec{a}, \vec{b}\right) = \left\| \vec{a} - \vec{b} \right\|.$$

EXAMPLE

If $\vec{a} = (1, 9, 2)$ and $\vec{b} = (7, 7, 1)$, then $d\left(\vec{a}, \vec{b}\right) = \sqrt{41}$.

THEOREM

If V is an inner product space, $\vec{a}$, $\vec{b}$, and $\vec{c}$ are vectors in V, and L is a scalar, then:

a) $\left\| \vec{a} \right\| \geq 0$ $\qquad\qquad$ $d\left(\vec{a}, \vec{b}\right) \geq 0$

b) $\left\| \vec{a} \right\| = 0.$ if and only if $\quad d\left(\vec{a}, \vec{b}\right) = 0$, if and only if

$\vec{a} = \vec{0}$ $\qquad\qquad\qquad \vec{a} = \vec{b}$

c) $\left\| L\vec{a} \right\| = |L| \left\| \vec{a} \right\|$ $\qquad d\left(\vec{a}, \vec{b}\right) = d\left(\vec{b}, \vec{a}\right)$

d) $\left\| \vec{a} + \vec{b} \right\| \leq \left\| \vec{a} \right\| + \left\| \vec{b} \right\|$ $\qquad d\left(\vec{a}, \vec{b}\right) \leq d\left(\vec{a}, \vec{c}\right) + d\left(\vec{c}, \vec{b}\right)$

The angle between two vectors $\vec{a}$ and $\vec{b}$ in an inner product space V is defined as:

$$\theta = arc\ \cos\left(\frac{<\vec{a}, \vec{b}>}{\left\| \vec{a} \right\| \left\| \vec{b} \right\|} \right) \text{ where } 0 \leq \theta \leq \pi.$$

Two vectors $\vec{a}$ and $\vec{b}$ in an inner product space are orthogonal if $<\vec{a}, \vec{b}> = 0$. If $\vec{a}$ is orthogonal to each vector in a set of vectors X, then $\vec{a}$ is orthogonal to X.

EXAMPLE

The vectors $(1, 2, -3)$ and $(-3, 9, 5)$ are orthogonal.

$1(-3) + 2(9) + (-3)5 = -3 + 18 - 15 = 0$

THEOREM

If $\vec{a}$ and $\vec{b}$ are orthogonal vectors in an inner product space, then:

$$\left\| \vec{a} + \vec{b} \right\|^2 = \left\| \vec{a} \right\|^2 + \left\| \vec{b} \right\|^2.$$

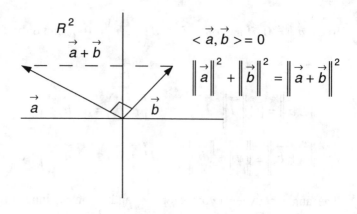

Figure 3.9

Problem Solving Examples:

 Show that the inner product can be derived from the theorem of Pythagoras and the law of cosines.

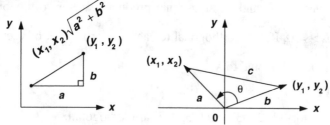

| **Figure 3.10** | **Figure 3.11** |

A   The Pythagorean theorem is used to derive the notion of distance between two points in the plane while the law of cosines enables angle measurements to be made.

Definition: $d(\alpha, \beta) = \sqrt{\alpha^2 \beta^2}$, the distance between the origin and the point (α, β).

Notice that $a = |x_1 - y_1|$ and $b = |x_2 - y_2|$.

From Figure 3.10, the distance between points (x_1, x_2) and (y_1, y_2) is

$$\left((x_1 - y_1)^2 + (x_2 - y_2)^2\right)^{\frac{1}{2}} = d\big((x_1 - y_1), (x_2 - y_2)\big).$$

So, any distance of R^2 can be expressed in terms of the function d. In Figure 3.11, the distance of (x_1, x_2), from the origin is given by $a = d(x_1, x_2)$ while the distance of (y_1, y_2) from 0 is given by $b = d(y_1, y_2)$. The distance from (x_1, x_2) to (y_1, y_2) is given by c. Then, the law of cosines states that

$$\cos \theta = \frac{a^2 + b^2 - c^2}{2ab} \qquad (1)$$

Since $a^2 + b^2 - c^2 = x_1^2 + x_2^2 + y_1^2 + y_2^2 - (x_1 - y_1)^2 - (x_2 - y_2)^2$
$$= 2(x_1 y_1 + x_2 y_2),$$

(1) may be rewritten as

$$\cos \theta = \frac{x_1 y_1 + x_2 y_2}{d(x_1, x_2) d(y_1, y_2)} \qquad (2)$$

The inner product of two vectors $\overrightarrow{u} = (x_1, x_2)$ and $\overrightarrow{u} = (x_1, x_2)$ is defined as $< \overrightarrow{u}, \overrightarrow{v} > = x_1 y_1 + x_2 y_2$. Thus,

$$\cos \theta = \frac{< u, v >}{d(x_1, x_2) d(y_1, y_2)}$$

Now notice that

$$d(x_1, x_2) = \sqrt{x_1^2 + x_2^2} = \sqrt{(x_1, x_2),(x_1, x_2)} = \sqrt{< \overrightarrow{u},, \overrightarrow{u}, >} = \left(< \overrightarrow{u}, \overrightarrow{u}, >\right)^{\frac{1}{2}}$$

Similarly, $d(y_1, y_2) = < \overrightarrow{u}, \overrightarrow{v} >^{\frac{1}{2}}$. Hence,

$$\cos \theta = \frac{< \overrightarrow{u}, \overrightarrow{v} >}{\left(< \overrightarrow{u}, \overrightarrow{u} >\right)^{\frac{1}{2}} \left(< \overrightarrow{v}, \overrightarrow{v} >\right)^{\frac{1}{2}}}. \qquad (3)$$

From (2), the inner product can be expressed in terms of the angle between two vectors and the distance between them:

$$< \vec{u}, \vec{v} > = x_1y_1 + x_2y_2 = (\cos \theta)\big(d(x_1, x_2)d(y_1, y_2)\big).$$

From (3), the length or norm of a vector can be expressed through the inner product.

 Find a vector orthogonal to $A = (2, 1, -1)$ and $B = (1, 2, 1)$.

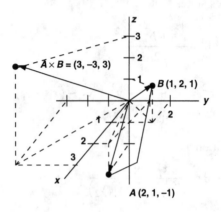

A ⨯ B = (3, −3, 3)
B (1, 2, 1)
A (2, 1, −1)

Figure 3.12

 A vector $V = (v_1, v_2, v_3)$ is said to be orthogonal to A and B if $A \times V = 0$ and $B \times V = 0$. In other words,

$$(v_1, v_2, v_3) \times (2, 1, -1) = 2v_1 + v_2 - v_3 = 0 \qquad (1)$$

and

$$(v_1, v_2, v_3) \times (1, 2, 1) = v_1 + 2v_2 + v_3 = 0 \qquad (2)$$

Adding (1) and (2)

$$3v_1 + 3v_2 = 0.$$

Therefore $v_1 = -v_2$. Let $v_2 = -3$. From $v_1 = -v_2$ and $v_3 = -(v_1 + 2v_2)$, the result is $v_1 = 3$ and $v_3 = -(3 - 6) = 3$. Therefore, the vector $(3, -3, 3)$ is orthogonal to $A = (2, 1, -1)$ and $B = (1, 2, 1)$.

Consider the vector space $C[0, 1]$ of all continuous functions defined on $[0, 1]$. If $f \in C[0, 1]$, show that

$$\left(\int_0^1 f^2(x)dx\right)^{\frac{1}{2}}$$

defines a norm on all elements of this vector space.

 Since f is continuous, f^2 is continuous. If a function is continuous, it is integrable; thus,

$$\left(\int_0^1 f^2(x)dx\right)^{\frac{1}{2}} = \|f\|$$

is well defined.

It is necessary to show that $\left(\int_0^1 f^2(x)dx\right)^{\frac{1}{2}}$ satisfies conditions a) – d) and a) – b) from page 112. If $f(x) \in C[0, 1]$, then $f^2(x) \geq 0$ on $[0, 1]$.

But this implies that $\left(\int_0^1 f^2(x)dx\right)^{\frac{1}{2}} \geq 0$. If $\|f\| = \left(\int_0^1 f^2(x)dx\right)^{\frac{1}{2}} = 0$,

then $f^2(x) = 0$.

This suggests that $f(x) = 0$ on $[0, 1]$ since a nonnegative function continuous over the interval $[0, 1]$ can have zero integral over $[0, 1]$ only if that function is identically zero on $[0, 1]$. So, $\|f\| = 0$ if and only if $f = 0$.

c) Let L be a scalar from the field over which $C[0, 1]$ is defined. Then,

$$\| Lf \| \int_0^1 \left[[Lf(x)]^2 dx \right]^{\frac{1}{2}} = |L| \left[\int_0^1 f^2(x)dx \right]^{\frac{1}{2}} = |L| \| f \|.$$

d) First, show that $\| f + g \|, f, g, \in C[0, 1]$ is well-defined. Now

$$\| f + g \| = \int_0^1 \left([f(x) + g(x)]^2 dx \right)^{\frac{1}{2}}$$

exists since sums and squares of continuous functions in $C[0, 1]$ are also in $C[0, 1]$ and, therefore, integrable over $[0, 1]$. Note also that $fg = \frac{1}{2}[(f+g)^2 - f^2 - g^2]$ is also in $C[0, 1]$ and integrable over $[0, 1]$.

It is now possible to show $\| f + g \| \leq \| f \| + \| g \|$.

$$\| f + g \| \leq \| f \| + \| g \| \Leftrightarrow \| f + g \|^2 \leq \left(\| f \| + \| g \| \right)^2 \qquad (1)$$

But,

$$\| f + g \|^2 = \int_0^1 [f(x) + g(x)]^2 dx \qquad (2)$$
$$= \int_0^1 f^2(x)dx + 2\int_0^1 f(x)g(x)dx + \int_0^1 g^2(x)dx$$

and

$$\left(\| f \| + \| g \| \right)^2 = \| f \|^2 + 2\| f \| \| g \| + \| g \|^2 \qquad (3)$$
$$= \int_0^1 f^2(x)dx + 2\left(\int_0^1 f^2(x)dx \right)^{\frac{1}{2}} \left(\int_0^1 g^2(x)dx \right)^{\frac{1}{2}}$$
$$+ \int_0^1 g^2(x)dx.$$

Comparing (2) and (3), we see that (1) can hold if and only if

$$\int_0^1 f(x)g(x)dx \leq \left(\int_0^1 f^2(x)dx \right)^{\frac{1}{2}} \left(\int_0^1 g^2(x)dx \right)^{\frac{1}{2}} \qquad (4)$$

By the properties of the absolute value functions,

$$\int_0^1 f(x)g(x)dx \leq \int_0^1 |f(x)| \times |g(x)|dx.$$

It can be proved that $\int_0^1 |f(x)| \times |g(x)|dx \leq \| f \| \| g \|$.

Let λ be a real variable and form $\int_0^1 \left[| f(x)| + \lambda |g(x)| \right]^2 dx \geq 0$. But,

$$\int_0^1 \left[| f(x)| + \lambda |g(x)| \right]^2 dx$$
$$= \lambda^2 \int_0^1 g^2(x)dx + 2\lambda \int_0^1 |f(x)||g(x)|dx + \int_0^1 f^2(x)dx$$

is a nonnegative quadratic polynomial in λ. Hence, it has no real roots and its discriminant is nonpositive; i.e.,

$$4\left[\int_0^1 |f(x)| |g(x)|dx \right]^2 - 4\int_0^1 g^2(x)dx \int_0^1 f^2(x)dx \leq 0$$

which implies

$$\int_0^1 |f(x)||g(x)|dx \le \left(\int_0^1 f^2(x)dx\right)^{\frac{1}{2}}\left(\int_0^1 g^2(x)dx\right)^{\frac{1}{2}},$$

as was to be shown.

Thus, $\left(\int_0^1 f^2(x)dx\right)^{\frac{1}{2}}$ defines a norm on $C[0,\,1]$ known as the Euclidean norm.

Q Let $C[0,\,1]$ be the vector space of all real-valued continuous functions defined on $[0,\,1]$. Consider the polynomials $2x$ and $1-2x^2$. Show that they are orthogonal with respect to the Euclidean norm. Then find their lengths.

A The inner product $< f,g >= \int_0^1 f(x)g(x)dx$ makes the space $C[0,\,1]$ a Euclidean space. Using this inner product, it is possible to find the length or norm of a vector: $\|f\| = \left[\int_0^1 [f(x)]^2\,dx\right]^{\frac{1}{2}}$. This norm is analagous to the norm of a vector $u = (u_1,\,u_2,...,\,u_n)$ in R^n; $\|u\| = \left(\sum u_1^2\right)^{\frac{1}{2}}$. Two vectors are orthogonal if their inner product equals zero, i.e., $f(x)$ and $g(x)$ are orthogonal if $\int_0^1 f(x)g(x)dx = 0$. The inner product of $f(x) = 2x$ and $g(x) = 1 - 2x^2$

$$= \int_0^1 2x\left(1 - 2x^2\right)dx$$

$$= -4\int_0^1 x^3dx + 2\int_0^1 x\,dx$$

$$= -x^4\Big|_0^1 + x^2\Big|_0^1 = -1+1 = 0$$

Thus, the two given functions are orthogonal. The length of $2x$ is given by

$$\|2x\| =< 2x,2x >^{\frac{1}{2}}= \left[\int_0^1 4x^2 dx\right]^{\frac{1}{2}} = \left[\frac{4}{3}x^3\Big|_0^1\right]^{\frac{1}{2}} = \frac{2}{\sqrt{3}}.$$

Similarly,

$$\left\| 1 - 2x^2 \right\| = \left[\int_0^1 \left(4x^4 - 4x^2 + 1\right)dx \right]^{\frac{1}{2}} = \left[\frac{4}{5}x^5 - \frac{4}{3}x^3 + x \Big|_0^1 \right]^{\frac{1}{2}} = \sqrt{\frac{7}{15}}.$$

3.9   Orthonormal Bases

If, in a set of vectors, in an inner product space, all pairs of distinct vectors are orthogonal, then the set is called an orthogonal set.

An orthonormal set is an orthogonal set in which each vector has a norm of one.

The process of multiplying a non-zero vector by the reciprocal of its norm to make it orthonormal is called normalizing the vector.

THEOREM

If $V = \left\{ \vec{v}_1, \vec{v}_2, ..., \vec{v}_n \right\}$ is a normal basis for an inner product

space X, and $\vec{a} \in X$, then:

$$\vec{a} = <\vec{a}, \vec{v}_1> \vec{v}_1 + <\vec{a}, \vec{v}_2> \vec{v}_2 + ... + <\vec{a}, \vec{v}_n> \vec{v}_n.$$

THEOREM

If $V = \left\{ \vec{v}_1, \vec{v}_2, ..., \vec{v}_n \right\}$ is a non-zero, orthogonal set, in an inner

product space, then V is linearly independent.

If $V = \left\{ \vec{v}_1, \vec{v}_2, ..., \vec{v}_n \right\}$ is an orthonormal set of vectors in an inner

product space X, and if Y is the space spanned by V, then every

vector $\vec{a} \in X$ can be expressed in the form $\vec{a} = \vec{y}_1 + \vec{y}_1$, where $\vec{y}_1 \in$

Y is called the orthogonal projection of $\vec{a}$ on X. $\vec{y}_2$ is orthogonal to Y

and is called the component of $\vec{a}$ orthogonal to Y. $\vec{y}_1$ and $\vec{y}_2$ are

calculated as follows:

a) $\vec{y_1} = <\vec{a}, \vec{v_1}> \vec{v_1} + <\vec{a}, \vec{v_2}> \vec{v_2} + \ldots + <\vec{a}, \vec{v_n}> \vec{v_n}$

b) $\vec{y_2} = \vec{a} - <\vec{a}, \vec{v_1}> \vec{v_1} - <\vec{a}, \vec{v_2}> \vec{v_2} - \ldots - <\vec{a}, \vec{v_n}> \vec{v_n}$

THEOREM

Every non-zero, finite dimensional inner product space has a basis consisting of orthonormal vectors, called an orthonormal basis.

3.9.1 Gram-Schmidt Process

The Gram-Schmidt Process converts any basis $\left\{\vec{a_1}, \vec{a_2}, \ldots, \vec{a_n}\right\}$

into an orthonormal basis $\left\{\vec{v_1}, \vec{v_2}, \ldots, \vec{v_n}\right\}$.

a) Let $\vec{v_1} = \dfrac{\vec{a_1}}{\left\|\vec{a_1}\right\|}$. $\vec{v_1}$ will have a norm of one.

b) Construct $\vec{v_2} = \dfrac{\vec{a_2} - <\vec{a_2}, \vec{v_1}> \vec{v_1}}{\left\|\vec{a_2} - <\vec{a_2}, \vec{v_1}> \vec{v_1}\right\|}$

c) Construct $\vec{v_3} = \dfrac{\vec{a_3} - <\vec{a_3}, \vec{v_1}> \vec{v_1} - <\vec{a_3}, \vec{v_2}> \vec{v_2}}{\left\|\vec{a_3} - <\vec{a_3}, \vec{v_1}> \vec{v_1} - <\vec{a_3}, \vec{v_2}> \vec{v_2}\right\|}$

Continue until an orthonormal set of vectors $\left\{\vec{v_1}, \vec{v_2}, \ldots, \vec{v_n}\right\}$ is obtained. This will be the orthonormal basis.

Problem Solving Examples:

a) The vectors $\{(1, 0), (0, 1)\}$ form an orthonormal basis for R^2. Find another orthonormal basis for R^2.

b) The vectors $\vec{u_1} = (1, 1, 1, 1)$, $\vec{u_2} = (1, -1, 1, -1)$, and, $\vec{u_3} = (1, 2, -1, -2)$ are orthogonal. Orthonormalize them.

a) The problem is to find two perpendicular vectors (thus, linearly independent vectors) in R^2 that are of unit length.

Let (u_1, u_2), $(v_1, v_2) \in R^2$. Then, $< \vec{u}, \vec{v} > = 0$ gives $u_1 v_1 + u_2 v_2 = 0$ which implies $u_1 v_1 = -u_2 v_2$. Thus, all vectors $\vec{u}$ and $\vec{v}$ such that $u_1 = -u_2$ and $v_1 = v_2$ will be orthogonal. Choose the simplest vectors $(1 -1)$ and $(1, 1)$. Their lengths are $\|(1, -1)\|$ and $\|(1, 1)\|$ or,

$$\|(1, -1)\| = \left[(1, -1) \times (1, -1)\right]^{\frac{1}{2}} = \sqrt{2};$$
$$\|(1, 1)\| = \left[(1, 1) \times (1, 1)\right]^{\frac{1}{2}} = \sqrt{2}.$$

Therefore, we divide each vector by its magnitude to get a new vector pointing in the same direction but with magnitude one. An orthonormal basis for R^2 is, therefore,

$$\left\{ \left(\frac{1}{\sqrt{2}}, -\frac{1}{\sqrt{2}} \right), \left(\frac{1}{\sqrt{2}}, \frac{1}{\sqrt{2}} \right) \right\}.$$

This means that any vector $(x, y) \in R^2$ can be expressed as a linear combination of the above basis.

b) First check that the three vectors in R^4 are mutually perpendicular.

$< (1, 1, 1, 1), (1, -1, 1, -1) > = (1) - 1(1) + 1(1) - 1(1) = 0$
$< (1, 1, 1, 1), (1, 2, -1, -2) > = 1(1) + 1(2) + 1(-1) + 1(-2) = 0$
$< (1, -1, 1, -1), (1, 2, -1, -2) > = 1(1) - 1(2) + 1(-1) - (1)(-2) = 0$

The length of the vectors is given by the norm.

$$\|(1, 1, 1, 1)\| = \; < (1, 1, 1, 1), (1, 1, 1, 1) >^{\frac{1}{2}} = \sqrt{4} = 2$$

$$\|(1, -1, 1, -1)\| = \; < (1, -1, 1, -1), (1, -1, 1, -1) >^{\frac{1}{2}} = \sqrt{4} = 2$$

$$\|(1, 2, -1, -2)\| = \; < (1, 2, -1, -2), (1, 2, -1, 2) >^{\frac{1}{2}} = \sqrt{10}$$

Dividing each vector by its norm results in normalized vectors. Thus,

$$\left\{ \left(\frac{1}{2}, \frac{1}{2}, \frac{1}{2}, \frac{1}{2} \right), \left(\frac{1}{2}, -\frac{1}{2}, \frac{1}{2}, -\frac{1}{2} \right), \left(\frac{1}{\sqrt{10}}, \frac{2}{\sqrt{10}}, \frac{-1}{\sqrt{10}}, \frac{-2}{\sqrt{10}} \right) \right\}$$

is an orthonormal set.

Q Show that the vectors $f_1 = \left(\frac{1}{2}, \frac{\sqrt{3}}{2} \right)$, $f_2 = \left(\frac{\sqrt{3}}{2}, -\frac{1}{2} \right)$ form an orthonormal basis for E^2. Then find the coordinates of an arbitrary vector $(x_1, x_2) \in E^2$.

A First show that the set $\{f_1, f_2\}$ is a basis for E^2. Since E^2 has dimension two, any set containing two linearly independent vectors forms a basis for this space. Now, any set of orthonormal vectors is linearly independent. To see this, let $G = \{g_1, g_2, \ldots, g_n\}$ be orthonormal. Suppose

$$a_1 g_1 + a_2 g_2 + \ldots + a_n g_n = \sum_{j}^{n} = 1 a_j g_j = 0 \tag{1}$$

Take the inner product with respect to g_1 on both sides of equation (1).

$$< g_1, \sum_{j=1}^{n} a_j g_j > = \; < g_1, 0 >, \tag{2}$$

but $<v, 0> = v \times 0 = 0$ for any vector v. So, (2) becomes

$$< g_1, \sum_{j=1}^{n} a_j g_j > = 0. \tag{3}$$

However, by the properties of an inner product

$$(\text{i.e., } <v, b_1 u_1 + b_2 u_2> = b_1 <v, u_1> + b_2 <v, u_2>),$$

we obtain from (3),

$$< g_1, \sum_{j=1}^{n} a_j g_j > = \sum_{j=1}^{n} a_j < g_1, g_j > = 0. \qquad (4)$$

G is an orthogonal set, so

$$\sum_{s \neq t} < g_s, g_t > = 0$$

and G is normalized. Thus,

$$< g_1, g_1 >^{\frac{1}{2}} = \| g \| = 1.$$

This implies $<g_1, g_1> = 1$. Therefore, (4) can be written as

$$a_1 <g_1, g_1> + a_2 <g_1, g_2> + \ldots + a_n <g_1, g_n> = 0.$$

This yields

$$a_1(1) + a_2(0) + \ldots + a_n(0) = 0$$
$$a_1 = 0$$

Similarly, it is possible to find $a_i = 0$, $i = 2,\ldots, n$. Beginning with equation (1), $\sum a_j g_j = 0$, we obtain

$$< g_1, \sum a_j g_j > = < g_1, 0 > = 0.$$

Then,

$$\sum_{j=1}^{n} a_j < g_i, g_j > = a_1 < g_i, g_i > + \therefore + a_i < g_i, g_i > + \ldots + a_n < g_i, g_n >$$

$$= a_1(0) + \ldots + a_i(1) + \ldots + a_n(0) = 0$$

which implies that $a_i = 0$. So,

$$\sum_{i=n}^{n} a_i g_i = 0 \neq a_i = 0, i = 1,\ldots,n$$

implies G is a set of linearly independent vectors.

Hence, if it is shown that f_1 and f_2 are orthogonal to each other and are normalized, $\{f_1, f_2\}$ will be a basis for E^2.

$$< f_1, f_2 > = < \left(\frac{1}{2}, \frac{\sqrt{3}}{2}\right), \left(\frac{\sqrt{3}}{2}, -\frac{1}{2}\right) > = \frac{\sqrt{3}}{4} - \frac{\sqrt{3}}{4} = 0.$$

Thus, f_1 and f_2 are mutually orthogonal. The norm of a vector $u = (u_1, u_2, \ldots, u_n)$ in E^n is given by:

$$\|u\| = \sum_{i=1}^{n} u_i^2$$

$$\|f_1\| = \left\| \left(\frac{1}{2}, \frac{\sqrt{3}}{2}\right) \right\| = \sqrt{\frac{1}{4} + \frac{3}{4}} = 1$$

$$\|f_2\| = \left\| \left(\frac{\sqrt{3}}{2}, -\frac{1}{2}\right) \right\| = \sqrt{\frac{3}{4} + \frac{1}{4}} = 1$$

Thus, f_1, f_2 are both of unit length, and $\{f_1, f_2\}$ forms a basis of E^2. Let $(x_1, x_2) \in E^2$. Then,

$$(x_1, x_2) = c_1 f_1 + c_2 f_2$$

$$x_1 = \frac{c_1}{2} + \frac{\sqrt{3}c_2}{2}; \ x_2 = \frac{\sqrt{3}c_1}{2} - \frac{c_2}{2}.$$

Solving this system for c_1, c_2.

$$c_1 = \frac{x_1 + \sqrt{3}x_2}{2}, \ c_2 = \frac{\sqrt{3}x_1 - x_2}{2}$$

and

$$(x_1, x_2) = \frac{\left(x_1 + \sqrt{3}x_2\right)f_1}{2} + \frac{\left(x_1\sqrt{3} - x_2\right)f_2}{2}.$$

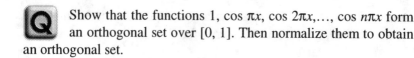

Q Show that the functions 1, cos πx, cos $2\pi x$,..., cos $n\pi x$ form an orthogonal set over [0, 1]. Then normalize them to obtain an orthogonal set.

A Consider the vector space $V = C[0, 1]$ of all real-valued continuous functions defined on the interval [0, 1]. The functions 1, cos πx, cos $2\pi x$,..., cos $n\pi x$ belong to this space. In order to define perpendicularity between elements of this space, it is necessary to define an inner product on $C[0, 1]$.

A suitable inner product on $C[0, 1]$ is

$$< f, g >= \int_0^1 f(x)g(x)dx.$$

To show that the given functions are pairwise orthogonal, it must be shown:

$$< 1, \cos m\pi x > = 0;$$
$$< \cos m\pi x, \cos n\pi x > = 0, m \neq n,$$

m an integer, n an integer, Now,

$$< 1, \cos m\pi x > = \int_0^1 \cos m\pi t \, dt$$

$$= \frac{1}{m\pi} \sin m\pi t \Big|_0^1 = 0.$$

Since $<f, g> = 0$ implies f and g are orthogonal, the constant function 1 is orthogonal to all the other members in the set.

$$< \cos m\pi x, \cos n\pi x > = \int_0^1 \cos m\pi x \cos n\pi t \, dt. \qquad (1)$$

But, $\cos m\pi x, \cos n\pi x = \frac{1}{2}\left[\cos(m + n)\pi x + \cos(m - n)\pi x\right]$. Hence, (1) becomes

$$\frac{1}{2}\int_0^1 \cos(m + n)\pi t \, dt + \frac{1}{2}\int_0^1 \cos(m - n)\pi t \, dt$$

$$= \frac{\sin(m + n)\pi t}{2(m + n)\pi}\Big|_0^1 + \frac{\sin(m - n)\pi t}{2(m - n)\pi}\Big|_0^1 = 0$$

Thus, the functions 1, $\cos \pi x$, $\cos 2\pi x$,..., $\cos n\pi x$, ... form an orthogonal set. Next, to normalize this set, find the length of each of the functions. The norm of a function f in $C[0, 1]$ is given by:

$$\| f \| = \left(\int_0^1 f^2(x)dx \right)^{\frac{1}{2}} = < f, f >^{\frac{1}{2}} .$$

$$\| 1 \| = < 1, 1 >^{\frac{1}{2}} = \left(\int_0^1 dt \right)^{\frac{1}{2}} = 1.$$

The function $f(x) = 1$ has norm 1.

$$\| \cos m\pi x \| = < \cos m\pi x, \cos m\pi x >^{\frac{1}{2}} = \left(\int_0^1 \cos^2 m\pi t \, dt \right)^{\frac{1}{2}}. \qquad (2)$$

By the half-angle formula, $\cos^2 m\pi x = \frac{1}{2} + \frac{1}{2} \cos 2m\pi x$. Hence, (2) becomes:

$$\left(\frac{1}{2} \int_0^1 \cos 2m\pi t + 1 \, dt \right)^{\frac{1}{2}} = \left[\frac{1}{2} \left(\frac{\sin 2m\pi t}{2m\pi} + t \right) \Big|_0^1 \right]^{\frac{1}{2}} = \frac{1}{\sqrt{2}}$$

Thus, the norm of $\cos m\pi x$ is $\frac{1}{\sqrt{2}}$. Since this is true for arbitrary integral values of m, the vectors $\sqrt{2} \cos m\pi x$ have unit norm. Thus, the functions 1, $\sqrt{2} \cos \pi x$, $\sqrt{2} \cos 2\pi x$,..., $\sqrt{2} \cos n\pi x$ form an orthonormal basis.

Let $F = \{(1, 0, 0, 1, 0), (0, 1, 1, -1, 0), (1, 1, 1, 1, 1)\}$ be a set of linearly independent vectors in E^5. Construct an orthogonal set G from F.

First show that the vectors in F are not orthogonal.

$$<(1, 0, 0, 1, 0), (0, 1, 1-1, 0)> = -1 \neq 0$$
$$<(1, 0, 0, 1, 0), (1, 1, 1, 1, 1)> = 2 \neq 0$$
$$<(0, 1, 1, -1, 0), (1, 1, 1, 1, 1)> = -1 \neq 0$$

Given a set of linearly independent vectors, the Gram-Schmidt procedure tells us how to obtain an orthonormal set.

Let

$$g_1 = f_1 = (1, 0, 0, 1, 0).$$

Let

$$g_2 = f_2 - \frac{< f_2, f_1 >}{< f_1, f_1 >} f_1 = (0,1,1,-1,0) + \tfrac{1}{2}(1,0,0,1,0) = (\tfrac{1}{2},1,1,-\tfrac{1}{2},0).$$

$$g_3 = f_3 \frac{< f_3, g_2 >}{< g_2, g_2 >} g_2 - \frac{< f_3, g_1 >}{< g_1, g_1 >} g_1$$

$$= (1,1,1,1,1) - \frac{\frac{2}{5}}{\frac{5}{2}}\left(\frac{1}{2},1,1,-\frac{1}{2},0\right) - \frac{2}{2}(1,0,0,1,0)$$

$$= \left(-\frac{2}{5},\frac{1}{5},\frac{1}{5},\frac{2}{5},1\right)$$

Therefore, $G = \left\{(1,0,0,1,0),\left(\frac{1}{2},1,1,-\frac{1}{2},0\right),\left(-\frac{2}{5},\frac{1}{5},\frac{1}{5},\frac{2}{5},1\right)\right\}.$

The set G is orthogonal.

 Find an orthonormalizing sequence v_1, v_2, v_3 for the following set of vectors in E^4:

$$u_1 = (1, -1, 1, -1); u_2 = (5, 1, 1, 1); u_3 = (2, 3, 4, -1).$$

A First verify that the set $\{u_1, u_2, u_3\}$ is linearly independent. The equation $c_1 u_1 + c_2 u_2 + c_3 u_3 = 0$ determines a system of equations which has as its only solution $c_1 = c_2 = c_3 = 0$. Thus, u_1, u_2, u_3 form a linearly independent set. Now, use the Gram-Schmidt orthogonalization procedure.

Let:

$$v_1 = \frac{(1,-1,1,-1)}{\sqrt{4}} = \frac{1}{2}(1,-1,1,-1).$$

Next,

$$v_2 = \frac{w_2}{\|w_2\|}$$

where

$$w_2 = u_2 - <v_1 u_2> v_1$$

$$= (5,1,1,1) - 2\left(\frac{1}{2}, -\frac{1}{2}, \frac{1}{2}, -\frac{1}{2}\right)$$

$$= (4,2,0,2)$$

Since:

$$\|w_2\| = \sqrt{24}, v_2 = \frac{(4,2,0,2)}{\sqrt{4 \times 6}}$$

$$= \frac{1}{\sqrt{6}}(2,1,0,1).$$

Finally,

$$v_3 = \frac{w_3}{\|w_3\|}$$

where

$$w_3 = u_3 - <v_1, u_3> v_1 - <v_2, u_3> v_2$$

$$= (2,3,4,-1) - 2\left(\frac{1}{2}, -\frac{1}{2}, \frac{1}{2}, -\frac{1}{2}\right) - \frac{\sqrt{6}}{\sqrt{6}}(2,1,0,1)$$

$$= (-1,3,3,-1).$$

Since:

$$\|w_3\| = \sqrt{20},$$

$$v_3 = \frac{1}{\sqrt{20}}(-1,3,3,-1).$$

The set

$$\{v_1, v_2, v_3\} = \left\{ \left(\frac{1}{2}, -\frac{1}{2}, \frac{1}{2}, -\frac{1}{2}\right), \left(\frac{2}{\sqrt{6}}, \frac{1}{\sqrt{6}}, 0, \frac{1}{\sqrt{6}}\right), \left(\frac{-1}{\sqrt{20}}, \frac{3}{\sqrt{20}}, \frac{3}{\sqrt{20}}, \frac{-1}{\sqrt{20}}\right) \right\}$$

is a set of orthonormal vectors constructed from $\{u_1, u_2, u_3\}$.

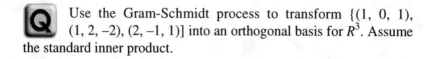

Q Use the Gram-Schmidt process to transform $\{(1, 0, 1),$ $(1, 2, -2), (2, -1, 1)]$ into an orthogonal basis for R^3. Assume the standard inner product.

A The Gram-Schmidt process constructs a set of orthogonal vectors from a given set of linearly independent vectors. Since a basis for R^3 must contain three linearly independent vectors, we check that the given vectors are linearly independent. Set $c_1(1, 0, 1)$ $+ c_2(1, 2, -2) + c_3(2, -1, 1) = 0$.

Thus,

$$c_1 + c_2 + 2c_3 = 0$$
$$2c_2 - c_3 = 0$$
$$c_1 - 2c_2 + c_3 = 0$$

The only solution to this system is $c_1 = c_2 = c_3 = 0$, i.e., the three vectors are linearly independent and form a basis for R^3.

Label the three vectors u_1, u_2, u_3, respectively. Choose u_1 and relabel it v_1. Thus, $v_1 = (1, 0, 1)$. The set now becomes $\{v_1, u_2, u_3\}$. We know u_1 and u_2 are independent, and the vector space they span is also spanned by $\{v_1, u_2\}$. We wish to construct a vector v_2 in this two-dimensional subspace of R^3 which is perpendicular to v_1. Choose v_2 as the candidate for this transfiguration. $v_2 \in Sp\{u_1, u_2\}$ implies $v_2 \in Sp\{v_1, u_2\}$ which, in turn, implies $v_2 = c_1 v_1 + c_2 u_2$ for some constants c_1 and c_2. The required orthogonality of v_1 and v_2 implies $<v_1, v_2> = 0$. Hence,

$$< v_1, c_1, v_1 + c_2 u_2 > = < v_1, c_1, v_1 > + < v_1, c_2, u_2 >$$
$$= c_1 < v_1, v_1 > + c_2 < v_1, u_2 >$$
$$= 0$$

Assume that $c_2 = 1$. Then,

$$c_1 = \frac{- < v_1, u_2 >}{< v_1, v_1 >} = \frac{- < v_1, u_2 >}{\| v_1 \|^2}$$

and hence,

$$v_2 = c_2 u_2 + c_1 u_1 = u_2 - \frac{<v_1, u_2> v_1}{\|v_1\|^2}$$

Thus, $v_2 = (1, 2, -2) \dfrac{- <(1,0,1), (1,2,-2)>}{\|(1,0,1)\|^2} (1, 0, 1)$

$$= (1, 2, -2) - \left(-\frac{1}{2}\right)(1, 0, 1)$$

$$= \left(\frac{3}{2}, 2, -\frac{3}{2}\right)$$

The set $\{v_1, v_2\}$ is orthogonal and forms a basis for the subspace of R^3.

Now seek a third vector v_3 that is orthogonal to both v_1 and v_2. Again, $Sp\{u_1, u_2, u_3\} = Sp\{v_1, v_2, v_3\}$, so $v_3 \in Sp\{u_1, u_2, u_3\}$ implies v_3 can be written as $v_3 = b_1 v_1 + b_2 v_2 + b_3 u_3$, subject to $<v_3, v_1> = <v_3, v_2> = 0$.

That is,

$$<b_1 v_1 + b_2 v_2 + b_3 u_3, v_1> = 0$$
$$<b_1 v_1 + b_2 v_2 + b_3 u_3, v_2> = 0$$

By the rules of the inner product,

$$b_1 <v_1, v_1> + b_2 <v_2, v_1> + b_3 <u_3, u_1> = 0$$
$$b_1 <v_1, v_2> + b_2 <v_2, v_2> + b_3 <u_3, v_2> = 0$$

Since $<v_2, v_1> = <v_1, v_2> = 0$, the above two equations reduce to

$$b_1 <v_1, v_1> + b_3 <u_3, v_1> = 0$$
$$b_2 <v_2, v_2> + b_3 <u_3, v_2> = 0$$

Assume $b_3 = 1$:

$$b_1 = \frac{- <u_3, v_1>}{\|v_1\|^2} \quad \text{and} \quad b_2 = \frac{- <u_3, u_2>}{\|v_2\|^2}.$$

Thus,

$$v_3 = b_3 u_3 + b_2 v_2 + b_1 v_1 = u_3 - \frac{<u_3, v_2> v_2}{\|v_2\|^2} - \frac{<u_3, v_1> v_1}{\|v_1\|^2}$$

$$= (2,-1,1) - \frac{<(2,-1,1),\left(\frac{3}{2},2,-\frac{3}{2}\right)>}{\frac{34}{4}}\left(\frac{3}{2},2,-\frac{3}{2}\right) - \frac{<(2,-1,1),(1,0,1)>}{2}(1,0,1)$$

$$= (2,-1,1) - \frac{\left(-\frac{1}{2}\right)}{\frac{34}{4}}\left(\frac{3}{2},2,-\frac{3}{2}\right) - \frac{3}{2}(1,0,1)$$

$$= (2,-1,1) + \left(\frac{3}{34},\frac{2}{17},-\frac{3}{34}\right) + \left(-\frac{3}{2},0,-\frac{3}{2}\right)$$

$$= \left(\frac{10}{17},-\frac{15}{17},-\frac{10}{17}\right).$$

The required orthogonal basis is:

$$\left\{(1,0,1),\left(\frac{3}{2},2,-\frac{3}{2}\right),\left(\frac{10}{17},-\frac{15}{17},-\frac{10}{17}\right)\right\}.$$

Any scalar multiple of these basis vectors will have the same properties of orthogonality, and so, if we wish, we may take {(1, 0, 1), (3, 4, −3), (2, −3, −2)} as an orthogonal basis of R^3.

3.10 Coordinates and Change of Basis

If $V = \left\{\vec{v_1}, \vec{v_2}, ..., \vec{v_n}\right\}$ is a basis for the vector space X, then

every vector $\vec{a} \in X$ can be expressed as $\vec{a} = c_1 \vec{v_1} = c_2 \vec{v_2} + ... + c_n \vec{v_n}$, where the scalars $c_1, c_2, ..., c_n$ are called the coordinates of X relative to V. The coordinate matrix is the $n \times 1$ matrix

$$\begin{bmatrix} c_1 \\ c_2 \\ \vdots \\ c_n \end{bmatrix} = \left(\vec{a}\right)_V.$$

THEOREM

If V is an orthonormal basis for an inner product space and if
$(\vec{X})_s = (x_1, x_2, \ldots, x_n)$ and $(\vec{Y})_s = (y_1, y_2, \ldots, y_n)$ then:

a) $\left\| \vec{x} \right\| = \left(x_1^2 + x_2^2 + \ldots + x_n^2 \right)^{\frac{1}{2}}$

b) $d\left(\vec{X}, \vec{Y} \right) = \left((x_1 - y_1)^2 (x_2 - y_2)^2 + \ldots + (x_n - y_n)^2 \right)^{\frac{1}{2}}$

c) $< \vec{X}, \vec{Y} > = x_1 y_1 + x_2 y_2 + \ldots + x_n y_n$

If the basis for a vector space V is being changed from

$$W = \left\{ \vec{w_1}, \vec{w_2}, \ldots, \vec{w_n} \right\} \text{ to } S = \left\{ \vec{s_1}, \vec{s_2}, \ldots, \vec{s_n} \right\},$$

then the transition matrix is defined as:

$$P = \left[\left(\vec{w_1} \right)_s : \left(\vec{w_2} \right)_s : \cdots : \left(\vec{w_n} \right)_s \right]$$

where $\left(\vec{w_j} \right)_s = \left(C_{1j}, C_{2j}, \ldots, C_{nj} \right)$

$j = 1, 2, \ldots, n$ s.t. $w_j = C_{1j}S_1 + C_{2j}S_2 + \ldots + C_{nj}S_n$

P is invertible, and P^{-1} is the transition matrix from S to W.

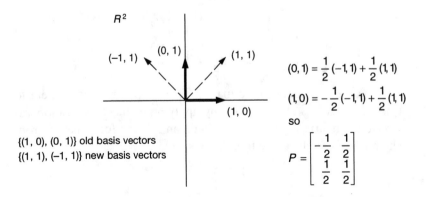

R^2

$(-1, 1)$ $(0, 1)$ $(1, 1)$

$(1, 0)$

{(1, 0), (0, 1)} old basis vectors
{(1, 1), (−1, 1)} new basis vectors

$(0, 1) = \dfrac{1}{2}(-1, 1) + \dfrac{1}{2}(1, 1)$

$(1, 0) = -\dfrac{1}{2}(-1, 1) + \dfrac{1}{2}(1, 1)$

so

$$P = \begin{bmatrix} -\dfrac{1}{2} & \dfrac{1}{2} \\ \dfrac{1}{2} & \dfrac{1}{2} \end{bmatrix}$$

Figure 3.13

THEOREM

If P is the transition matrix from one orthonormal basis to another, then $P^{-1} = P^t$.

The square matrix A with the property $A^{-1} = A^t$ is called an orthogonal matrix.

THEOREM

If A is an $n \times n$ matrix, then the following are equivalent:

a) A is orthogonal.

b) The row and column vectors of A form an orthonormal set in R^n.

Problem Solving Examples

Let P_4 denote the vector space of all polynomials of degree at most equal to four. Let V be the subspace of P_4 spanned by $S = \{\alpha_1, \alpha_2, \alpha_3, \alpha_4\}$ where $\alpha_1 = t^4 + t^2 + 2t + 1$, $\alpha_2 = t^4 + t^2 + 2t + 2$, $\alpha_3 = 2t^4 + t^3 + t + 2$, and $\alpha_4 = t^4 + t^3 - t^2 - t$. Find a basis for V.

First, find a matrix representation of the vectors in the given problem. We can write the four given polynomials as

$$\begin{bmatrix} 1 & 0 & 1 & 2 & 1 \\ 1 & 0 & 1 & 2 & 2 \\ 2 & 1 & 0 & 1 & 2 \\ 1 & 1 & -1 & -1 & 0 \end{bmatrix} \begin{bmatrix} t^4 \\ t^3 \\ t^2 \\ t \end{bmatrix} = \begin{bmatrix} \alpha_1 \\ \alpha_2 \\ \alpha_3 \\ \alpha_4 \end{bmatrix} \qquad (1)$$

Let V be the row space of the matrix. Then, if B is row equivalent to A, its row space is also V. Now, by applying elementary row operations to the matrix in (1), we can obtain an echelon form matrix from which a basis for V can be found. Thus,

$$A = \begin{bmatrix} 1 & 0 & 1 & 2 & 1 \\ 1 & 0 & 1 & 2 & 2 \\ 2 & 1 & 0 & 1 & 2 \\ 1 & 1 & -1 & -1 & 0 \end{bmatrix}.$$

Replace row 2 by row 2 + (–row 1).
Replace row 3 by row 3 + (–2 row 1).
Replace row 4 by row 4 + (–row 1).

$$\begin{bmatrix} 1 & 0 & 1 & 2 & 1 \\ 0 & 0 & 0 & 0 & 1 \\ 0 & 1 & -2 & -3 & 0 \\ 0 & 1 & -2 & -3 & -1 \end{bmatrix}$$

Replace row 4 by row 3 + (–row 4).

$$\begin{bmatrix} 1 & 0 & 1 & 2 & 1 \\ 0 & 0 & 0 & 0 & 1 \\ 0 & 1 & -2 & -3 & 0 \\ 0 & 1 & 0 & 0 & 1 \end{bmatrix}$$

Replace row 2 by row 3.
Replace row 3 by row 2.
Replace row 4 by row 4 + (–row 2).

Therefore, an equivalent matrix is:

$$B = \begin{bmatrix} 1 & 0 & 1 & 2 & 0 \\ 0 & 1 & -2 & -3 & 0 \\ 0 & 0 & 0 & 0 & 1 \\ 0 & 0 & 0 & 0 & 0 \end{bmatrix}. \tag{2}$$

Hence, a basis for the row space of A is $\{\beta_1, \beta_2, \beta_3\}$, where $\beta_1 = [1, 0, 1, 2, 0]$, $\beta_2 = [0, 1, -2, -3, 0]$, and $\beta_3 = [0, 0, 0, 0, 1]$. Multiplying (2) by the standard basis for P_4,

$$\begin{bmatrix} 1 & 0 & 1 & 2 & 0 \\ 0 & 1 & -2 & -3 & 0 \\ 0 & 0 & 0 & 0 & 1 \\ 0 & 0 & 0 & 0 & 0 \end{bmatrix} \begin{bmatrix} t^4 \\ t^3 \\ t^2 \\ t \end{bmatrix}$$

we see that a basis for the subspace V of P_4 is:

$$\{t^4 + t^2 + 2t, \ t^3 - 2t^2 - 3t, \ 1\}.$$

Let U and W be the following subspaces of R^4.

$$U = \{[a, b, c, d] : b + c + d = 0\},$$
$$W = \{[a, b, c, d] : a + b = 0, c = 2d\}$$

Find the dimension and a basis of: a) U, b) W, c) $U \cap W$.

A Recall that family $\{u_1,\ldots, u_k\}$ of vectors in a vector space U is a basis for U if: (a) it is linearly independent and (b) it spans U. A vector space U that has a basis consisting of a finite number of vectors is said to be finite-dimensional. The number of vectors in such a basis is called the dimension of U.

a) We seek a basis for the set of solutions (a, b, c, d) to the equation

$$b + c + d = 0 \text{ or } 0\,(a) + b + c + d = 0.$$

Choose the free variables to be a, c, and d. Set (1) $a = 1$, $c = 0$, $d = 0$; (2) $a = 0$, $c = 1$, $d = 0$; and (3) $a = 0$, $c = 0$, $d = 1$ to obtain the respective solutions.

$$u_1 = (1, 0, 0, 0),\ u_2 = (0, -1, 1, 0),\ u_3 = (0, -1, 0, 1)$$

The set $\{u_1, u_2, u_3\}$ is a basis of U, and *dim* $U = 3$. Examine more closely what has been done.

The equation $b + c + d = 0$ is equivalent to the relationship $0\,(a) + b + c + d = 0$ between the variables a, b, c, and d. We can choose "a" and two others freely and solve for the remaining variable. Choose b to be our dependent variable. Any vector $(a, b = c - d, c, d)$ lies in the subspace, and any set of three linearly independent vectors of this form will span the subspace. The vectors given by $(1, b, 0, 0)$, $(0, b, 1, 0)$, and $(0, b, 0, 1)$ must be linearly independent and, thus, immediately give us a basis for the subspace when we solve for b.

b) We seek a basis for the set of solutions (a, b, c, d) to the system

$$
\begin{array}{ccc}
a + b = 0 & \text{or} & a + b = 0 \\
c = 2d & & c - 2d = 0
\end{array}
$$

Choose the free variables to be b and d. Set (1) $b = 1$, $d = 0$; (2) $b = 0$, $d = 1$ to obtain the respective solutions

$$u_1 = (-1, 1, 0, 0), \ u_2 = (0, 0, 2, 1)$$

The set $\{u_1, u_2\}$ is a basis of W and $dim \ W = 2$.

c) $U = W$ consists of those vectors (a, b, c, d) which satisfy the conditions defining U and the conditions defining W, i.e., the three equations

$$b + c + d = 0 \qquad\qquad a + b = 0$$
$$a + b = 0 \qquad \text{or} \qquad b + c + d = 0$$
$$c = 2d \qquad\qquad c - 2d = 0$$

Choose the free variable to be d. Set $d = 1$ to obtain the solution $u = (3, -3, 2, 1)$. Thus, $\{u\}$ is a basis of $U \cap W$, and $dim \ (U \cap W) = 1$.

 Let $B_1 = \{(2, 1), (1, 0)\}$ and $B_2 = \{(-1, 2), (3, 1)\}$ be two bases for R^2, and let the matrix of $T(x, y)$ with respect to B_1

be $A = \begin{bmatrix} 2 & 0 \\ 1 & 3 \end{bmatrix}$.

What is the matrix of $T(x, y)$ with respect to B_2?

A Since the matrix of a transformation depends on the basis chosen, a change in basis will result in a change in the matrix of the transformation.

Let (u_1, u_2), (v_1, v_2) be two bases for R^2. Then $T : R^2 \to R^2$ is represented by two matrices determined by the relations:

$$T(u_1) = a_{11}u_1 + a_{12}u_2$$
$$T(u_2) = a_{21}u_1 + a_{22}u_2 \qquad\qquad (1)$$

$$T(v_1) = b_{11}v_1 + b_{12}v_2$$
$$T(v_2) = b_{21}v_1 + b_{22}v_2 \qquad\qquad (2)$$

From (1) and (2) we obtain the matrices

$$A = \begin{bmatrix} a_{11} & a_{21} \\ a_{12} & a_{22} \end{bmatrix} \qquad B = \begin{bmatrix} b_{11} & b_{21} \\ b_{12} & b_{22} \end{bmatrix}.$$

The question is: How can we get from A to B? To do so, the transition matrix from u_i to v_i and the transition matrix from v_i to u_i are needed. Since $v_1, v_2 \in R^2$,

$$v_1 = p_{11}u_1 + p_{12}u_2$$
$$v_2 = p_{21}u_1 + p_{22}u_2 \qquad (3)$$

Since $u_1, u_2 \in R^2$,

$$u_1 = q_{11}v_1 + q_{12}v_2$$
$$u_2 = q_{21}v_1 + q_{22}v_2. \qquad (4)$$

Thus,

$$P = \begin{bmatrix} p_{11} & p_{21} \\ p_{12} & p_{22} \end{bmatrix} \quad \text{and} \quad Q = \begin{bmatrix} q_{11} & q_{21} \\ q_{12} & q_{22} \end{bmatrix}$$

are the required matrices. According to a theorem in linear algebra, the matrix product $QAP = B$, i.e., it represents T with respect to the basis $\{v_1, v_2\}$. Examining the given problem, we are given A, the matrix of T with respect to B_1. Thus, we must find the transition matrices.

$$(-1, 2) = p_{11}(2, 1) + p_{12}(1, 0) = 2(2, 1) - 5(1, 0)$$
$$(3, 1) = p_{21}(2, 1) + p_{22}(1, 0) = 1(2, 1) + 1(1, 0)$$

Hence, $P = \begin{bmatrix} 2 & 1 \\ -5 & 1 \end{bmatrix}$. Similarly,

$$(2, 1) = q_{11}(-1, 2) + q_{12}(3, 1) = \frac{1}{7}(-1, 2) + \frac{5}{7}(3, 1)$$
$$(1, 0) = q_{21}(-1, 2) + q_{22}(3, 1) = -\frac{1}{7}(-1, 2) + \frac{2}{7}(3, 1)$$

Hence $Q = \begin{bmatrix} \frac{1}{7} & -\frac{1}{7} \\ \frac{5}{7} & \frac{2}{7} \end{bmatrix}$. Then,

$$QAP = \begin{bmatrix} \frac{1}{7} & -\frac{1}{7} \\ \frac{5}{7} & \frac{2}{7} \end{bmatrix} \begin{bmatrix} 2 & 0 \\ 1 & 3 \end{bmatrix} \begin{bmatrix} 2 & 1 \\ -5 & 1 \end{bmatrix}$$

$$= \begin{bmatrix} \frac{1}{7} & -\frac{1}{7} \\ \frac{5}{7} & \frac{2}{7} \end{bmatrix} \begin{bmatrix} 4 & 2 \\ -13 & 4 \end{bmatrix}$$

$$= \frac{1}{7} \begin{bmatrix} 1 & -1 \\ 5 & 2 \end{bmatrix} \begin{bmatrix} 4 & 2 \\ -13 & 4 \end{bmatrix}$$

$$= \frac{1}{7} \begin{bmatrix} 17 & -2 \\ -6 & 18 \end{bmatrix}$$

$$= B$$

Note that $Q = P^{-1}$ or $PQ = I$. Therefore, $B = P^{-1}AP$. Using this equation, the matrix B which represents T with respect to the new basis $\{v_1, v_2\}$ can also be obtained.

 The set $B = \{1, x\}$ is a basis for the vector space P^1 where P^1 is defined to be the vector space of all polynomials of degree less than or equal to 1 over the field of real numbers. Show that the coordinates of an arbitrary function in P^1, using the basis B, are unique.

A An arbitrary element of P^1 has the form: $f(x) = a + bx$ where a, b are elements of R (real numbers). Then $f(x) = (a)(1) + (b)(x)$.

Thus, $f(x)$ has been expressed as a linear combination of vectors in B. To show uniqueness, let $f(x) = c(1) + d(x)$ where c and d, both elements of R, are different from a and b, respectively. Certainly, $f(x) - f(x) = 0$ so,

$$a(1) + b(x) - [c(1) + d(x)]$$
$$= (a - c)1 + (b - d)x = 0$$

But the zero polynomial is the polynomial all of whose coefficients are zero. Therefore, $(a - c) = 0$ and $(b - d) = 0$ which implies that $a = c$ and $b = d$. Hence, $f(x)$ is uniquely represented by $(a)(1) + (b)(k)$.

Quiz: Vector Spaces

1. The following is the definition of the cross product:

   Let $U = u_1 i + u_2 j + u_3 k$ and $V = v_1 i + v_2 j + v_3 k$ where $\vec{i}, \vec{j},$ and $\vec{k}$ are the unit vectors $(1, 0, 0)$, $(0, 1, 0)$, and $(0, 0, 1)$, respectively. Then, their cross-product is the vector $\vec{U} \times \vec{V}$ defined by

   $$\vec{U} \times \vec{V} = (u_2 v_3 - u_3 v_2)\,\vec{i} + (u_3 v_1 - u_1 v_3)\,\vec{j} + (u_1 v_2 - u_2 v_1)\,\vec{k}.$$

   For ease of computation, it is convenient to view the cross-product as analogous to the determinant

   $$\vec{U} \times \vec{V} = \begin{bmatrix} \vec{i} & \vec{j} & \vec{k} \\ u_1 & u_2 & u_3 \\ v_1 & v_2 & v_3 \end{bmatrix}$$

   $$= \left(u_2 v_3 - u_3 v_2\right)\vec{i} - \left(u_1 v_3 - u_3 v_1\right)\vec{j} + \left(u_1 v_2 - u_2 v_1\right)\vec{k}$$

   Then the cross product $\vec{u} \times \vec{v}$ of the vectors

   $$\vec{u} = 2\vec{i} - \vec{j} + 3\vec{k}$$
   $$\vec{v} = \vec{i} - 2\vec{j} - \vec{k}$$

   is given by

   (A) $7\vec{i} + \vec{j} + 3\vec{k}.$ (D) $-5\vec{i} + 5\vec{j} + 5\vec{k}.$

   (B) $5\vec{i} + 5\vec{j} + 3\vec{k}.$ (E) $7\vec{i} - \vec{j}.$

   (C) $-3.$

2. Suppose $\vec{a}$ and $\vec{b}$ are vectors of R^2 and further, that $\vec{a}$ and $\vec{b}$ are linearly independent.

Let $\vec{c} = \dfrac{<\vec{a}, \vec{b}>}{\left\| \vec{a} \right\|^2} \vec{a} - \vec{b}$

   (A) $\vec{a}$ and $\vec{c}$ are linearly dependent.

   (B) $\vec{a}$ and $\vec{c}$ are orthogonal.

   (C) $\vec{b}$ and $\vec{c}$ are dependent.

   (D) $\vec{c}$ is the zero vector.

   (E) $\vec{b}$ and $\vec{c}$ are orthogonal.

3. The vector $(1, 2, 3)$ is orthogonal to which of the following vectors?

   (A) $(-4, -4, 4)$.   (D) $(-1, -2, -3)$.

   (B) $(-4, -4, -4)$.   (E) $(3, 2, 1)$.

   (C) $(4, 4, 4)$.

4. If V is an inner product space, $\vec{a}$ and $\vec{b}$ vectors in V, and d the distance between $\vec{a}$ and $\vec{b}$, which one of the following statements is *not* true?

   (A) $\| a \| \geq 0$.

   (B) $\| a \| = 0$ if and only if $a = 0$.

   (C) $d(a, b) \geq 0$.

   (D) $d(a, b) + d(b, a) = 0$.

   (E) $\| a + a \| \leq \| a \| + \| b \|$.

5. Which of the following sets of vectors forms a basis for the vector space spanned by the set of vectors $(4, 2, 3, -2)$, $(-15, -3, 6, 12)$, and $(-5, 7, -21, -2)$?

    (A) $(4, 2, 3, -2)$, $(-15, -3, 6, 12)$, and $(-5, 7, -21, -2)$.

    (B) $(-15, -3, 6, 12)$ and $(15, 3, -6, -12)$.

    (C) $(4, 2, 3, -2)$ and $(15, 3, -6, -12)$.

    (D) $(4, 2, 3, -2)$ and $(-15, -3, -6, -12)$.

    (E) $(1, 2, 5, -2)$, $(-15, -3, 6, 12)$, and $(-5, -7, -31, -2)$.

6. The vector $(-4, 1, -3)$ is *not* a linear combination of the vector $(1, 2, 3)$ and which one of the following vectors?

    (A) $(-2, -1, -3)$.   (D) $(-5, -1, -6)$.

    (B) $(-7, -5, -12)$.   (E) $(1, 11, 12)$.

    (C) $(-6, -9, -9)$.

7. Let $\vec{u} = u_1 \vec{i} + u_2 \vec{j} + u_3 \vec{k}$. The curl of $\vec{u}$, denoted curl $(\vec{u})$, is the cross product of the vector operator

$$\nabla \equiv \vec{i} \frac{\partial}{\partial x} + \vec{j} \frac{\partial}{\partial y} + \vec{k} \frac{\partial}{\partial z}$$

and the vector $\vec{u}$:

$$\text{curl}(\vec{u}) = \nabla \times \vec{u} = \begin{vmatrix} \vec{i} & \vec{j} & \vec{k} \\ \frac{\partial}{\partial x} & \frac{\partial}{\partial y} & \frac{\partial}{\partial z} \\ u_1 & u_2 & u_3 \end{vmatrix}$$

Find the curl of $\vec{u} = xyz \, \vec{i} + xy^2 \, \vec{j} + yz \, \vec{k}$.

(A) $xy\,\vec{i} + (x - yz)\,\vec{j} + 2y\,\vec{k}$.

(B) $(x - z)\,\vec{i} - yz\,\vec{j} + xyz\,\vec{k}$.

(C) $z\,\vec{i} + xy\,\vec{j} + (y^2 - xz)\,\vec{k}$.

(D) $xy\,\vec{i} - (z - y)\,\vec{j} + (xy - yz)\,\vec{k}$.

(E) $(xy - yz)\,\vec{i} - yz\,\vec{j} + x\,\vec{k}$.

8. Let V be the vector space of functions $f : R \to R$. Let S be the subspace generated by $\{e^x, e^{2x}, e^{-2x}\}$. Define D_x to be the derivative operator on S. Find the determinant of D_x.

(A) 2.   (D) 1.

(B) 0.   (E) −1.

(C) −4.

9. The column space of a 5×6 matrix is spanned by the vectors $(1, 0, 0, 0, 0)$, $(0, 0, 1, 0, 0)$, and $(2, 0, 3, 0, 0)$. Find the dimension of the solution space of the matrix.

(A) 3.   (D) 2.

(B) 4.   (E) 5.

(C) 6.

10. For the inner product $< A, B > = \text{trace}\,(B^t A)$ defined on the vector space of 2 by 2 matrices on R, find the square of the norm of

$$T = \begin{bmatrix} 1 & 3 \\ 2 & -1 \end{bmatrix}.$$

(A) 5.   (D) 25.

(B) 10.   (E) 49.

(C) 15.

ANSWER KEY

1.	(D)	6.	(C)
2.	(B)	7.	(C)
3.	(A)	8.	(C)
4.	(D)	9.	(B)
5.	(C)	10.	(C)

CHAPTER 4

Linear Transformations

4.1 Linear Transformations

If V and X are vector spaces and f is a function that relates a vector in V with a unique vector in X, then f is said to map X into V; $T : X \to V$.

If T is a function that associates a vector $\vec{a}$ with a vector $\vec{b}$, then $\vec{a}$ is the image of $\vec{b}$ under T.

If T is a function mapping the vector space X into the vector space Y, then T is called a linear transformation if:

a) $T\left(\vec{a} + \vec{b}\right) = T\left(\vec{a}\right) + T\left(\vec{b}\right)$, for all $\vec{a}, \vec{b} \in X$

b) $T\left(L\,\vec{a}\right) = LT\left(\vec{a}\right)$ for all $\vec{a} \in X$, and scalars L.

If A is an $m \times n$ matrix and B is an $n \times 1$ matrix, then the matrix transformation function $T(B)$ is defined as:

$$T(B) = AB$$

EXAMPLE

If $A = \begin{bmatrix} 2 & 3 \\ 4 & 1 \end{bmatrix}$ and $B = \begin{bmatrix} 1 \\ 2 \end{bmatrix}$, then

$$T(B) = AB$$

$$= \begin{bmatrix} 8 \\ 6 \end{bmatrix}.$$

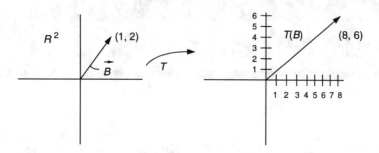

Figure 4.1

If V is a vector space, then the function $T\left(\vec{a}\right) = \vec{0}$ for all $\vec{a} \in V$ is called the zero transformation.

If V is a vector space, then the function $T\left(\vec{a}\right) = \vec{a}$ for every $\vec{a} \in V$ is called the identity transformation.

If V is a vector space and L is a scalar, then the function $T\left(\vec{a}\right) = L\vec{a}$ for every $\vec{a} \in V$ is called a dilation of V if $L > 1$, or a contraction of V if $0 < L < 1$.

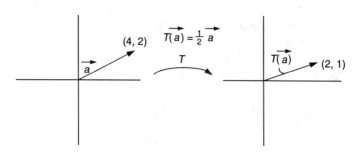

Figure 4.2 A contraction

If V is an inner product space, and if $\left\{\vec{x}_1, \vec{x}_2, \ldots, \vec{x}_n\right\}$ is an or-

thonormal basis of X, a finite dimensional subspace of V, then the function

$$T\left(\vec{a}\right) = <\vec{a}_1, \vec{x}_1> \vec{x}_1 + <\vec{a}_1, \vec{x}_2> \vec{x}_2 + \ldots + <\vec{a}_1, \vec{x}_n> \vec{x}_n$$

for every $\vec{a} \in V$ is called the orthogonal projection of V into X.

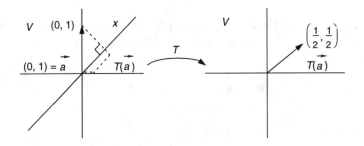

Figure 4.3 Let the basis for X be $\left\{\left(\dfrac{1}{2}, \dfrac{1}{2}\right)\right\}$ and let T be

an orthogonal projection.

Problem Solving Examples:

  Find the linear transformation that corresponds to the 2×3 matrix

$$\begin{bmatrix} 1 & 1 & 1 \\ 1 & 0 & 1 \end{bmatrix}.$$

There is a correspondence between 2×3 matrices and linear transformations from R^3 to R^2. To see this, let $B_3 = \{(1, 0, 0),$ $(0, 1, 0), (0, 0, 1)\}$ be the standard basis for R^3, and let $T(x_1, x_2, x_3)$ $= (a_{11}x_1 + a_{12}x_2 + a_{13}x_3, a_{21}x_1 + a_{22}x_2 + a_{23}x_3)$ be a linear transformation from R^3 to R^2. Then, $T(1, 0, 0) = (a_{11}, a_{21})$, $T(0, 1, 0) = (a_{12}, a_{22})$, $T(0, 0, 1) = (a_{13}, a_{23})$.

Let B_2 be the standard basis for $R^2 : B_2 = \{(1, 0), (0, 1)\}$. Then:

$$T(1,0,0) = a_{11}(1,0) + a_{21}(0,1)$$
$$T(0,1,0) = a_{12}(1,0) + a_{22}(0,1)$$
$$T(0,0,1) = a_{13}(1,0) + a_{23}(0,1)$$

Defining $x = (1, 0, 0)$; $y = (0, 1, 0)$; $z = (0, 0, 1)$; $u = (1, 0)$; $v = (0, 1)$:

$$T\begin{bmatrix} x \\ y \\ z \end{bmatrix} = \begin{bmatrix} a_{11} & a_{21} \\ a_{12} & a_{22} \\ a_{13} & a_{23} \end{bmatrix} \begin{bmatrix} u \\ v \end{bmatrix}. \tag{1}$$

The transpose of the above matrix is the matrix of the transformation (with respect to the standard basis). Thus, the matrix of the linear transformation is

$$A = \begin{bmatrix} a_{11} & a_{12} & a_{13} \\ a_{21} & a_{22} & a_{23} \end{bmatrix}.$$

Now consider the given matrix

$$A = \begin{bmatrix} 1 & 1 & 1 \\ 1 & 0 & 1 \end{bmatrix}.$$

Hence, $a_{11} = 1$, $a_{12} = 1$, $a_{13} = 1$

$$a_{21} = 1, a_{22} = 0, a_{23} = 1$$

Therefore, the linear transformation in R^3 corresponding to the given matrix is $T(x_1, x_2, x_3) = T(x_1 + x_2 + x_3, x_1 + x_3)$.

 What are the effects of the following transformations on points (x_1, x_2) in the plane? The transformations are represented by

$$A = \begin{bmatrix} a_{11} & a_{12} \\ a_{21} & a_{22} \end{bmatrix}.$$

a) $\quad A = \begin{bmatrix} 1 & 0 \\ 0 & -1 \end{bmatrix}$ b) $\quad A = \begin{bmatrix} -1 & 0 \\ 0 & -1 \end{bmatrix}$

c) $\quad A = \begin{bmatrix} 0 & -1 \\ 1 & 0 \end{bmatrix}$ d) $\quad A = \begin{bmatrix} 1 & 0 \\ 0 & 0 \end{bmatrix}$

e) $A = \begin{bmatrix} 2 & 0 \\ 0 & 0 \end{bmatrix}$ f) $A = \begin{bmatrix} 1 & 1 \\ 1 & 1 \end{bmatrix}$

g) $A = \begin{bmatrix} 2 & 0 \\ 0 & 2 \end{bmatrix}$ h) $A = \begin{bmatrix} 2 & 0 \\ 0 & 3 \end{bmatrix}$

i) $A = \begin{bmatrix} 1 & 1 \\ 0 & 1 \end{bmatrix}$

A A transformation is a mapping from a vector space V to a vector space W. Any linear transformation can be represented by a matrix. Each of the given transformations is linear.

$A: [x_1, x_2] \rightarrow AX$ has the following interpretation.

By the rules of matrix multiplication,

$$AX = \begin{bmatrix} a_{11} & a_{12} \\ a_{21} & a_{22} \end{bmatrix} \begin{bmatrix} x_1 \\ x_2 \end{bmatrix}' = \begin{bmatrix} a_{11}x_1 + 2_{12}x_2 \\ a_{21}x_1 + a_{22}x_2 \end{bmatrix} \qquad (1)$$

a) Using (1), $A: [x_1, x_2] \rightarrow \begin{bmatrix} 1 & 0 \\ 0 & -1 \end{bmatrix} \begin{bmatrix} x_1 \\ x_2 \end{bmatrix}' = \begin{bmatrix} x_1 \\ -x_2 \end{bmatrix}$.

Thus, a point with coordinates (x_1, x_2) is sent to a point whose first coordinate is the same, but whose second coordinate is the negative of the original coordinate.

From Figure 4.4, we see that only points on the x_1 axis are unaltered by the transformation. All other points are reflected across the x_1 axis.

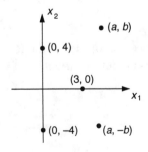

Figure 4.4

b) $AX = \begin{bmatrix} -1 & 0 \\ 0 & -1 \end{bmatrix} \begin{bmatrix} x_1 \\ x_2 \end{bmatrix} = \begin{bmatrix} -x_1 \\ -x_2 \end{bmatrix}.$

The point (a, b) goes to $(-a, -b)$. The only invariant point is the origin $(0, 0)$. The effect on the points geometrically is to shift the vectors they represent in R^2 by an angle of $180°$.

c) $AX = \begin{bmatrix} 0 & -1 \\ 1 & 0 \end{bmatrix} \begin{bmatrix} x_1 \\ x_2 \end{bmatrix} = \begin{bmatrix} -x_2 \\ x_1 \end{bmatrix}.$

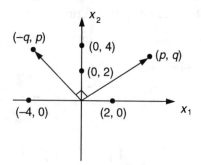

Figure 4.5

From Figure 4.5, we see the transformation is a rotation about the origin through a right angle in the counterclockwise direction. The origin is the only invariant point.

d) $AX = \begin{bmatrix} 1 & 0 \\ 0 & 0 \end{bmatrix} \begin{bmatrix} x_1 \\ x_2 \end{bmatrix} = \begin{bmatrix} x_1 \\ 0 \end{bmatrix}.$

Every point in the plane is projected orthogonally onto the x_1-axis. All points on the x_1-axis are left invariant.

e) $\quad AX = \begin{bmatrix} 2 & 0 \\ 0 & 0 \end{bmatrix} \begin{bmatrix} x_1 \\ x_2 \end{bmatrix} = \begin{bmatrix} 2x_1 \\ 0 \end{bmatrix}.$

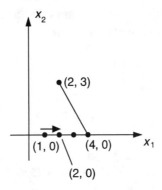

Figure 4.6

Points on the x_1-axis have their distance from the origin doubled. Other points are projected onto the x_1-axis. The x_1-axis is an invariant line, i.e., the line is unchanged by the transformation although individual points are displaced on it.

f) $\quad AX = \begin{bmatrix} 1 & 1 \\ 1 & 1 \end{bmatrix} \begin{bmatrix} x_1 \\ x_2 \end{bmatrix} = \begin{bmatrix} x_1 + x_2 \\ x_1 + x_2 \end{bmatrix}.$

Here, the result of the transformation is that all points are sent to the $x_1 = x_2$ line. In other words, the transformation sends the vector space R^2 into the real line R^1 which, in the $x_1\ x_2$ plane, is represented by $x_1 = x_2$. Vectors on the line $x_1 = x_2$ have their magnitudes doubled.

g) $\quad AX = \begin{bmatrix} 2 & 0 \\ 0 & 2 \end{bmatrix} \begin{bmatrix} x_1 \\ x_2 \end{bmatrix} = \begin{bmatrix} 2x_1 \\ 2x_2 \end{bmatrix}.$

The distance of every point from the origin is doubled. The origin is the only invariant point, but every line through the origin is an invariant line.

h) $AX = \begin{bmatrix} 2 & 0 \\ 0 & 3 \end{bmatrix} \begin{bmatrix} x_1 \\ x_2 \end{bmatrix} = \begin{bmatrix} 2x_1 \\ 3x_2 \end{bmatrix}.$

The points are stretched; however, the magnitude of the stretch is not the same in all directions. The origin is the only invariant point, but both the x_1 and x_2 axes are invariant lines.

i) $AX = \begin{bmatrix} 1 & 1 \\ 0 & 1 \end{bmatrix} \begin{bmatrix} x_1 \\ x_2 \end{bmatrix} = \begin{bmatrix} x_1 + x_2 \\ x_2 \end{bmatrix}.$

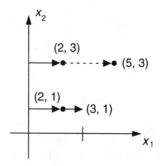

Figure 4.7

Points on the x_1-axis are left invariant. Other points are slid along horizontally, the amount of slide being proportional to the distance from the x_1-axis.

 Find the inverses of the following transformations using the matrices associated with the transformations:

1) T is a counterclockwise rotation in R^2 through the angle θ.

2) T is reflection in the y-axis in R^2.

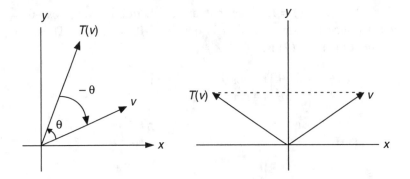

Figure 4.8   **Figure 4.9**

 A linear transformation in R^2 is described by its effects on the basis vectors $B = \{(1, 0), (0, 1)\}$.

Letting $v_1 = (0, 1)$, $T(1, 0) = (\cos \theta, \sin \theta)$. If $v_2 = (0, 1)$,

$$T(0, 1) = \left(\cos \theta + \frac{\pi}{2}, \sin \theta + \frac{\pi}{2} \right).$$

Since $\cos\left(\theta + \frac{\pi}{2} \right) = -\sin \theta$ and $\sin\left(\theta + \frac{\pi}{2} \right) = \cos \theta$,

$$T(0, 1) = (-\sin \theta, \cos \theta).$$

Thus,

$$T(x, y) = \begin{bmatrix} \cos \theta & -\sin \theta \\ \sin \theta & \cos \theta \end{bmatrix} \begin{bmatrix} x \\ y \end{bmatrix}. \tag{1}$$

From Figure 4.8, the inverse transformation is a clockwise rotation through θ or a counterclockwise rotation through $-\theta$. Hence, its matrix, using (1) is

$$S(x, y) = \begin{bmatrix} \cos(-\theta) & -\sin(-\theta) \\ \sin(-\theta) & \cos(-\theta) \end{bmatrix} \begin{bmatrix} x \\ y \end{bmatrix}. \tag{2}$$

To check that (2) is indeed the required inverse, we must show

$TS(x, y) = ST(x, y) = I$. Since the cosine function is even, $\cos(-\theta) = \cos\theta$. Since the sine function is odd, $\sin(-\theta) = -\sin\theta$. Thus, the product of (1) and (2) is:

$$
\begin{bmatrix} \cos\theta & -\sin\theta \\ \sin\theta & \cos\theta \end{bmatrix} \begin{bmatrix} \cos\theta & \sin\theta \\ -\sin\theta & \cos\theta \end{bmatrix} \begin{bmatrix} x \\ y \end{bmatrix}
$$

$$
= \begin{bmatrix} \cos^2\theta + \sin^2\theta & \cos\theta\sin\theta - \cos\theta\sin\theta \\ \sin\theta\;\cos\theta - \cos\theta\;\sin\theta & \sin^2\theta + \cos^2\theta \end{bmatrix} \begin{bmatrix} x \\ y \end{bmatrix}
$$

$$
= \begin{bmatrix} 1 & 0 \\ 0 & 1 \end{bmatrix} \begin{bmatrix} x \\ y \end{bmatrix} = \begin{bmatrix} x \\ y \end{bmatrix}.
$$

Similarly, $TS(x, y) = I$.

From Figure 4.9, we see that (x, y) is sent to $(-x, y)$. We can find the matrix associated with this transformation by describing its effects on the basis vectors $B = \{(1, 0), (0, 1)\}$. Now $T(1, 0) = (-1, 0)$ and $T(0, 1) = (0, 1)$.

Hence,

$$
T(x, y) = \begin{bmatrix} -1 & 0 \\ 0 & 1 \end{bmatrix} \begin{bmatrix} x \\ y \end{bmatrix}. \tag{3}
$$

Looking again at Figure 4.9, we suspect that the inverse of T will be given by another reflection in the y-axis, i.e., the transformation is its own inverse. If so, the matrix in (3) is its own inverse. To check,

$$
\begin{bmatrix} -1 & 0 \\ 0 & 1 \end{bmatrix} \begin{bmatrix} -1 & 0 \\ 0 & 1 \end{bmatrix} \begin{bmatrix} x \\ y \end{bmatrix} = \begin{bmatrix} 1 & 0 \\ 0 & 1 \end{bmatrix} \begin{bmatrix} x \\ y \end{bmatrix} = \begin{bmatrix} x \\ y \end{bmatrix}.
$$

4.2   Kernel and Range

THEOREM:

If $T : A \to B$ is a linear transformation with $\vec{a} \in A$ and $\vec{b} \in B$, then:

a) $T\left(\vec{0}\right) = \vec{0}$

b) $T\left(-\vec{a}\right) = -T\left(\vec{a}\right)$

c) $T\left(\vec{a} - \vec{b}\right) = T\left(\vec{a}\right) - T\left(\vec{b}\right)$

The kernel (or null space) of a linear transformation T ($ker(T)$) is that set of vectors which T maps into $\vec{0}$.

EXAMPLE

If $T : A \to B$ is a zero transformation, then the vector space A is the kernel of T.

$$ker\,(T) = \left\{\vec{a} \in A \;\middle|\; T(a) = \vec{0}\right\} = T^{-1}\left(\vec{0}\right)$$

The range of linear transformation $T : A \to B$, ($R(T)$), is the set of vectors in B that are images under T of at least one vector in A.

$$Im\,(T) = \text{RANGE}\,(T) = \left\{\vec{b} \in B \;\; \vec{a} \in A \mid T(a) = b\right\}$$

EXAMPLE

If $T : A \to B$ is a zero transformation, then $R(T) = \vec{0}$.

Let $T = \begin{pmatrix} -1 & 1 \\ 1 & -1 \end{pmatrix}$

Then $ker(T) = \{(x, y) \mid x = y\}$

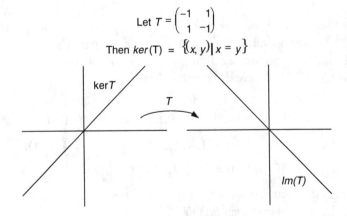

Figure 4.10

If $T : R^n \to R^m$ represents multiplication by an $m \times n$ matrix A, then $ker(T)$ is the solution space of A; $R(T)$ is the column space of A.

If T is a linear transformation, then $dim(R(T))$ is called the rank of T, and $dim(ker(T))$ is called the nullity of T.

THEOREM

If A is an n-dimensional vector space and $T : A \to B$ is a linear transformation, then

$$(\text{rank of } T) + (\text{nullity of } T) = n.$$

THEOREM

If A is an $m \times n$ matrix, then the dimensions of the solution space of $A \vec{X} = \vec{0}$ equal n-(rank of A).

Problem Solving Examples:

Let $T: R^4 \to R^3$ be a linear transformation defined by

$$T(x, y, z, t) = (x - y + z + t, \ x + 2z - t, \ x + y + 3z - 3t).$$

Find a basis and the dimension of the

a) image of T \qquad\qquad b) kernel of T

a) We know that if e_i ($i = 1,..., n$) spans R^n and A is any linear operator, then $A(e_i)$ ($i = 1,..., n$) spans the image space of A ($Im\ (A)$). So we see first what effect T has on a set of basis vectors in R^4. Choosing the standard basis:

$$\{(1, 0, 0, 0), (0, 1, 0, 0,), (0, 0, 1, 0), (0, 0, 0, 1)\},$$

$$T(1, 0, 0, 0) = (1, 1, 1); T(0, 1, 0, 0) = (-1, 0, 1);$$

$$T(0, 0, 1, 0) = (1, 2, 3); T(0, 0, 0, 1) = (1, -1, -3).$$

To form the matrix representing the linear transformation, we take as rows the vectors that span $Im\ (T)$.

$$\begin{bmatrix} 1 & 1 & 1 \\ -1 & 0 & 1 \\ 1 & 2 & 3 \\ 1 & -1 & -3 \end{bmatrix}$$

We use elementary row operations on (1) to row reduce it to echelon form.

$$\begin{bmatrix} 1 & 1 & 1 \\ 0 & 1 & 2 \\ 0 & 1 & 2 \\ 0 & -2 & -4 \end{bmatrix} \rightarrow \begin{bmatrix} 1 & 1 & 1 \\ 0 & 1 & 2 \\ 0 & 0 & 0 \\ 0 & 0 & 0 \end{bmatrix}$$

Thus, there are two independent vectors in the spanning set, and the dimension of $Im\ (T)$ is equal to two. A basis for $Im\ (T)$ is $\{(1, 1, 2), (0, 1, 2)\}$.

To find a basis for $ker\ (T)$, we argue as follows: $ker\ (T)$, by definition, contains the vectors (x, y, z, t) such that $T(x, y, z, t) = (0, 0, 0)$. Hence,

$$\begin{aligned} x - y + z + t &= 0 \\ x + 2z - t &= 0 \\ x + y + 3z - 3t &= 0 \end{aligned} \quad (2)$$

Write the matrix of coefficients of (2) and reduce to echelon form (by row operations).

$$\begin{matrix} R_1 : \\ R_2 : \\ R_3 : \end{matrix} \begin{pmatrix} 1 & -1 & 1 & 1 \\ 1 & 0 & 2 & -1 \\ 1 & 1 & 3 & -3 \end{pmatrix}$$

$$\begin{matrix} R_1 = R_1' : \\ R_2 - R_1 = R_2' : \\ R_3 - R_1 = R_3' : \end{matrix} \begin{pmatrix} 1 & -1 & 1 & 1 \\ 0 & 1 & 1 & -2 \\ 0 & 2 & 2 & -4 \end{pmatrix}$$

$$\begin{matrix} R_2' + R_1' = R_1' \\ R_2' = R_2' \\ (-2)R_2' + R_3' = R_3' \end{matrix} \begin{pmatrix} 1 & 0 & 2 & -1 \\ 0 & 1 & 1 & -2 \\ 0 & 0 & 0 & 0 \end{pmatrix} \quad (3)$$

(3) has two independent rows so $dim(ker(T)) = 2$. In this case, $dim(ker(T))$ is equal to $dim(Im\ (T))$. From (3), a basis for $ker(T)$ is $\{(1, -1, 1, 1), (0, 1, 1, -2)\}$.

 Let $T: R^4 \to R^4$ be the linear transformation defined by:

$$T\begin{bmatrix} x_1 \\ x_2 \\ x_3 \\ x_4 \end{bmatrix} = \begin{bmatrix} 0 \\ x_1 + x_2 \\ x_4 \\ 0 \end{bmatrix}.$$

Find the dimensions of $Im(T)$ and $ker(T)$.

 Let $T: V \to W$ be a linear transformation. Then, according to an important theorem in linear algebra,

$$dim\ (V) = dim\ (Im(T)) + dim(ker(T)$$

or

$$dim\ (ker\ (T)). \tag{1}$$

The dimension of V for the given transformation is $dim(R^4) = 4$. Hence, we must find either $dim(Im\ (T))$ or $dim(ker\ (T))$, since then we can solve (1) to find the other. Now a linear transformation is fully described by its effects on the basis vectors. Taking the standard basis of R^4,

$$T\begin{bmatrix} 1 \\ 0 \\ 0 \\ 0 \end{bmatrix} = \begin{bmatrix} 0 \\ 1 \\ 0 \\ 0 \end{bmatrix} v_1; \quad T\begin{bmatrix} 0 \\ 1 \\ 0 \\ 0 \end{bmatrix} = \begin{bmatrix} 0 \\ 1 \\ 0 \\ 0 \end{bmatrix} v_2;$$

$$T\begin{bmatrix} 0 \\ 0 \\ 1 \\ 0 \end{bmatrix} = \begin{bmatrix} 0 \\ 0 \\ 0 \\ 0 \end{bmatrix} = v_3; \quad T\begin{bmatrix} 0 \\ 0 \\ 0 \\ 1 \end{bmatrix} = \begin{bmatrix} 0 \\ 0 \\ 1 \\ 0 \end{bmatrix} = v_4$$

From these vectors, we can construct a set that spans the image of T. We know $\{v_i\}\ i = 1, 2, 3, 4$ spans $Im\ (T)$ so $dim(Im\ (T))$ is the number of independent v_i's. We eliminate v_3 since the 0-vector is dependent on any vector and v_2 since $v_1 = v_2$ (they are dependent).

Thus,

$$S = \left\{ \begin{bmatrix} 0 \\ 1 \\ 0 \\ 0 \end{bmatrix} \begin{bmatrix} 0 \\ 0 \\ 1 \\ 0 \end{bmatrix} \right\}$$

spans T. Furthermore, the two vectors in S are linearly dependent, i.e., they form a basis. Thus, $Im(T)$ has dimension equal to two. From (1) $dim\ (ker(T))$ also equals two and we have

$$dim\ (Im)\ (T) = 2$$
$$dim\ (ker)\ (T) = 2$$

4.3   Linear Transformations from Rⁿ to Rᵐ

For every linear transformation $T : R^n \rightarrow R^m$, with $\left\{ \vec{v}_1, \vec{v}_2, \ldots, \vec{v}_n \right\}$

being the standard basis for R^n, there is an $m \times n$ matrix A, called the standard matrix for T, for which T is a multiple of A. A has

$\left\{ T\left(\vec{v}_1\right), T\left(\vec{v}_2\right), \ldots, T\left(\vec{v}_n\right) \right\}$ as its column vectors:

$$A = \left[T\left(\vec{v}_1\right) \vdots T\left(\vec{v}_2\right) \vdots \ldots \vdots T\left(\vec{v}_n\right) \right]$$

EXAMPLE

If $B = \begin{bmatrix} x \\ y \\ z \end{bmatrix}$, then for $T(A) = \begin{bmatrix} x - z \\ z + y \\ y + z \\ y \end{bmatrix}$

the standard matrix is

$$A = \begin{bmatrix} 1 & 0 & -1 \\ 1 & 1 & 0 \\ 0 & 1 & 1 \\ 0 & 1 & 0 \end{bmatrix}.$$

The standard matrix for a matrix transformation is the matrix itself.

Problem Solving Examples:

Let $T: R^2 \rightarrow R^2$ be given by

$$T(x_1, x_2) = (4x_1 - 2x_2, 2x_1 + x_2)$$

and let $\{(1, 1), (-1, 0)\}$ be a basis for R^2. Compute the matrix of T in the given basis.

We are given

$$T(1, 1) = (4 - 2, 2 + 1) = (2, 3)$$
$$T(-1, 0) = (-4 - 0, -2 + 0) = (-4, -2)$$

In terms of the basis $\{(1, 1), (-1, 0)\}$, $(2, 3) = a_{11} (1, 1) + a_{12}(-1, 0)$

or

$$2 = a_{11} - a_{12}$$
$$3 = a_{11}$$

Hence, $a_{11} = 3$, $a_{12} = 1$ and $(2, 3) = 3(1, 1) + 1(-1, 0)$. Similarly, $(-4, -2) = a_{21} (1, 1) + a_{22} (-1, 0)$; hence, $a_{21} = -2$, $a_{22} = 2$. Thus, $(-4, -2) = (-2(1, 1) + 2(-1, 0)$. The matrix of T relative to the given basis is, therefore,

$$A = \begin{bmatrix} 3 & -2 \\ 1 & 2 \end{bmatrix}$$

Show that the transformation $T(x_1, x_2) = (3x_1, -x_2, -x_1 + x_2)$ which has the matrix

$$\begin{bmatrix} 3 & 0 \\ 0 & -1 \\ -1 & 2 \end{bmatrix}$$

can also be considered as a transformation

$$(x_1, x_2)T = (3x_1, -x_2, -x_1 + x_2).$$

with matrix

$$\begin{bmatrix} 3 & 0 & -1 \\ 0 & -1 & 2 \end{bmatrix}.$$

 This problem illustrates the ambiguous notation prevelant in current texts on linear algebra.

One approach to describing the correspondence between linear transformations and matrices is the following:

Let $T: R^n \to R^m$ be a linear transformation. Choose the standard bases for R^n and R^m, i.e., $\{e_i\}$ and $\{e_j\}$, $i = 1,\ldots, n$; $j = 1,\ldots, m$ where e_i and e_j represent the ith and jth n and m-tuples with 1 in the ith and jth coordinate and zeros everywhere else. Then,

$$T(e_1) = a_{11}e_1 + a_{12}e_2 + \ldots + a_{1m}e_m$$
$$T(e_2) = a_{21}e_1 + a_{22}e_2 + \ldots + a_{2m}e_m$$
$$\vdots \tag{1}$$
$$T(e_n) = a_{n1}e_1 + a_{n2}e_2 + \ldots + a_{nm}e_m$$

System (1) may be expressed in matrix form as:

$$T\begin{bmatrix} e_1 \\ e_2 \\ \cdot \\ \cdot \\ \cdot \\ e_n \end{bmatrix} = \begin{bmatrix} a_{11} & a_{12} & \cdots & a_{1m} \\ a_{21} & a_{22} & \cdots & a_{2m} \\ \cdot & & & \cdot \\ \cdot & & & \cdot \\ \cdot & & & \cdot \\ a_{n1} & a_{n2} & \cdots & a_{nm} \end{bmatrix}\begin{bmatrix} e_1 \\ e_2 \\ \cdot \\ \cdot \\ \cdot \\ e_m \end{bmatrix} \tag{2}$$

The transpose of the matrix in (2) is called the matrix of the transformation. Thus,

$$A = \begin{bmatrix} a_{11} & a_{12} & \cdots & a_{n1} \\ a_{12} & a_{22} & \cdots & a_{n2} \\ \cdot & & & \cdot \\ \cdot & & & \cdot \\ \cdot & & & \cdot \\ a_{1m} & a_{2m} & \cdots & a_{nm} \end{bmatrix} \tag{3}$$

Why is the matrix in (2) transposed to obtain the matrix (3)? Note that now the operator T can be replaced by the matrix operator A. The operator A and the column of unit vectors of R^n are now conformable for multiplication. In (2) the matrix was of order $n \times m$ and could not be post-multiplied by an $n \times 1$ column vector.

Thus, a transformation $T: R^n \to R^m$ corresponds to an $m \times n$ matrix.

The second approach is as follows: Let $T: R^m \to R^n$ be given by $(x_1, x_2, \ldots)T$. Then, proceeding as in the first approach,

$$e_1 T = a_{11} e_1 + a_{12} e_2 + \ldots + a_{1m} e_m$$
$$e_2 T = a_{21} e_1 + a_{22} e_2 + \ldots + a_{2m} e_m$$
$$\vdots$$
$$e_n T = a_{n1} e_1 + a_{n2} e_2 + \ldots + a_{nm} e_m$$

Rewriting in matrix form,

$$[e_1 \ e_2 \ldots]T = \begin{bmatrix} a_{11} & a_{12} & \ldots & a_{1m} \\ a_{21} & a_{22} & \ldots & a_{2m} \\ \cdot & & & \\ \cdot & & \cdot & \\ \cdot & & & \\ \cdot & & \cdot & \\ a_{n1} & a_{n2} & \ldots & a_{nm} \end{bmatrix} \begin{bmatrix} e_1 \\ \cdot \\ \cdot \\ \cdot \\ \cdot \\ \cdot \\ e_m \end{bmatrix} \tag{4}$$

Now the $n \times m$ matrix in (4) can be used to represent the transformation T. Thus, a transformation $T: R^n \to R^m$ corresponds to an $n \times m$ matrix.

The two approaches are illustrated by the given transformation. First, $T: R^2 \to R^3$ is represented by a 3×2 matrix; then $T: R^2 \to R^3$ is represented by a 2×3 matrix. To check, we see by using matrix multiplication that:

$$T(x_1, x_2) = \begin{bmatrix} 3 & 0 \\ 0 & -1 \\ -1 & 2 \end{bmatrix} \begin{bmatrix} x_1 \\ x_2 \end{bmatrix} = [x_1, x_2] \begin{bmatrix} 3 & 0 & -1 \\ 0 & -1 & 2 \end{bmatrix}$$

$$= (x_1, x_2)T = (3x_1, -x_2, -x_1 + x_2)$$

4.4 Matrices of Linear Transformations

If A is an n-dimensional vector space with basis $S = \left\{ \vec{s_1}, \vec{s_2}, \ldots, \vec{s_n} \right\}$, and B is an m-dimensional vector space with basis $Z = \left\{ \vec{z_1}, \vec{z_2}, \ldots, \vec{z_n} \right\}$, then for the transformation $T : A \to B$, the matrix of T with respect to the bases S and Z is defined as:

$$\left[T\left(\vec{s_1}\right)_z \quad T\left(\vec{s_2}\right)_z . 3 \quad T\left(\vec{s_n}\right)_z \right]$$

$$TC\left(\vec{s_j}\right)_z = \begin{bmatrix} c_1 \\ \vdots \\ c_m \end{bmatrix} \text{ such that}$$

$$T\left(\vec{s_j}\right) = c_1 z_1 + c_2 z_2 + \ldots + c_m z_m$$

If V is a vector space with a finite basis S, then for $T : V \to V_1$ the matrix of T with respect to the basis S will be just the standard matrix for T.

Problem Solving Examples:

Q If $T : R^3 \to R^3$ is defined by the matrix

$$A = \begin{bmatrix} 2 & 3 & 1 \\ 1 & 2 & 3 \\ 3 & 1 & 2 \end{bmatrix}$$

find $(1, 1, -1)A$. What is $(x_1, x_2, x_3) \, T$? Use the standard basis.

 Here T is operating on vectors in R^3 from the right. The matrix A represents T with respect to the standard basis for R^3, $B = \{(1, 0, 0), (0, 1, 0)\ (0, 0, 1)\}$. Thus, it can replace T wherever T occurs and vice-versa. To find $(1, 1, -1)A$ we simply premultiply A by the given row vector.

$$(1,1,-1)A = [1,1,-1]\begin{bmatrix} 2 & 3 & 1 \\ 1 & 2 & 3 \\ 3 & 1 & 2 \end{bmatrix} = [0,4,2]$$

To find the T that corresponds to A, we premultiply A by (x_1, x_2, x_3), an arbitrary vector in R^3.

$$[x_1, x_2, x_3]\begin{bmatrix} 2 & 3 & 1 \\ 1 & 2 & 3 \\ 3 & 1 & 2 \end{bmatrix}$$

$$= [2x_1 + x_2 + 3x_3, 3x_1 + 2x_2 + x_3, x_1 + 3x_2 + 2x_3]$$

To veryify this, notice the following. Let T operate on the basis vectors for R^3; then, $T(1, 0, 0) = (2, 3, 1)$; $T(0, 1, 0) = (1, 2, 3)$, and $T(0, 0, 1) = (3, 1, 2)$. Thus, $(x_1, x_2, x_3)T$

$$= [2x_1 + x_2 + 3x_3, 3x_1 + 2x_2 + x_3, x_1 + 3x_2 + 2x_3].$$

Let $B: R^3 \to R^3$ be given by:

$$(x_1, x_2, x_3)B = (2x_1 + x_3, x_1 + x_2 - x_3, x_1 - x_2 + x_3).$$

The linear transformation B has a matrix representation with respect to the standard basis for R^3, $\{(1, 0, 0), (0, 1, 0), (0, 0, 1)\}$. What are the row vectors of this matrix?

Since any vector in R^3 can be written as a linear combination of the basis vectors, we need only study the effect of the transformation on the basis $\{(1, 0, 0), (0, 1, 0), (0, 0, 1)\}$.

$$(1, 0, 0)B = (2(1) + 0(0), 1(1) + 1(0) - 1(0), 1(1) - 1(0) + 1(0))$$
$$= (2, 1, 1)$$
$$(0, 1, 0)B = (0, 1, -1)$$
$$(0, 0, 1)B = (1, -1, 1)$$

Thus, the matrix of the transformation is

$$\begin{bmatrix} 2 & 1 & 1 \\ 0 & 1 & -1 \\ 1 & -1 & 1 \end{bmatrix}$$

and the row vectors are $(2, 1, 1)$, $(0, 1, -1)$, and $(1, -1, 1)$.

4.5 Similarity

Let V be a finite dimensional vector space and $T : V \to V$ its linear operator. If A is the matrix of T with respect to a basis X, and B is the matrix of T with respect to a basis Y, and if P is the transition matrix from X to Y, then $B = PAP^{-1}$.

If A and B are square matrices, then A and B are similar if an invertible matrix P exists such that $B = PAP^{-1}$.

Problem Solving Examples:

 Show that the matrices M' and M are similar, where

$$M = \begin{bmatrix} 1 & 1 \\ 0 & 1 \end{bmatrix} \text{ and } M' = \begin{bmatrix} 1 & 0 \\ 1 & 1 \end{bmatrix}.$$

 Two $n \times n$ matrices M and M' are similar if $M = PM' P^{-1}$, where P is an invertible $n \times n$ matrix.

To see that M and M' are similar, let

$$P = \begin{bmatrix} 0 & 1 \\ 1 & 0 \end{bmatrix}.$$

Then

$$P^{-1} = \begin{bmatrix} 0 & 1 \\ 1 & 0 \end{bmatrix} \text{ since } P^{-1} = \frac{1}{\det(P)} \text{ adj } P, \text{ and}$$

$$PM'P^{-1} = \begin{bmatrix} 0 & 1 \\ 1 & 0 \end{bmatrix} \begin{bmatrix} 1 & 0 \\ 1 & 1 \end{bmatrix} \begin{bmatrix} 0 & 1 \\ 1 & 0 \end{bmatrix}$$

$$= \begin{bmatrix} 0 & 1 \\ 1 & 0 \end{bmatrix} \begin{bmatrix} 0 & 1 \\ 1 & 1 \end{bmatrix}$$

$$= \begin{bmatrix} 1 & 1 \\ 0 & 1 \end{bmatrix} = M$$

 Given:

$$A = \begin{bmatrix} 1 & 1 \\ 0 & 1 \end{bmatrix}; \text{ and } P = \begin{bmatrix} 1 & 1 \\ 1 & -1 \end{bmatrix}.$$

(a) Find P^{-1}.

(b) Find $P^{-1}AP$.

(c) Verify that, if B is similar to A, then A is similar to B.

(d) Show that $B^k = P^{-1}A^kP$ if $B = P^{-1}AP$ where k is any positive integer.

 (a) It is known that $P^{-1} = \dfrac{1}{\det(P)}$ adj (P).

$$\det(P) = \begin{bmatrix} 1 & 1 \\ 1 & -1 \end{bmatrix} = -1 - 1 = -2$$

$$\text{adj}(P) = \begin{bmatrix} -1 & -1 \\ -1 & 1 \end{bmatrix}$$

Therefore,

$$P^{-1} = \begin{bmatrix} \frac{1}{2} & \frac{1}{2} \\ \frac{1}{2} & -\frac{1}{2} \end{bmatrix}.$$

(b) $P^{-1}AP = \begin{bmatrix} \frac{1}{2} & \frac{1}{2} \\ \frac{1}{2} & -\frac{1}{2} \end{bmatrix} \begin{bmatrix} 1 & 1 \\ 0 & 1 \end{bmatrix} \begin{bmatrix} 1 & 1 \\ 1 & -1 \end{bmatrix}$

$\qquad = \begin{bmatrix} \frac{1}{2} & \frac{1}{2} \\ \frac{1}{2} & -\frac{1}{2} \end{bmatrix} \begin{bmatrix} 2 & 0 \\ 1 & -1 \end{bmatrix}$

$\qquad = \begin{bmatrix} \frac{3}{2} & -\frac{1}{2} \\ \frac{1}{2} & \frac{1}{2} \end{bmatrix}$

We say that the matrix B is similar to the matrix A if there is an invertible matrix P such that

$$B = P^{-1}AP.$$

Therefore, let

$$B = \begin{bmatrix} \frac{3}{2} & -\frac{1}{2} \\ \frac{1}{2} & \frac{1}{2} \end{bmatrix}$$

and then B is similar to A.

(c) If P is invertible and $B = P^{-1}AP$. Then

$$PBP^{-1} = P\left(P^{-1}AP\right)P^{-1}$$

$$= \left(PP^{-1}\right)A\left(PP^{-1}\right)$$

$$= A \text{ since } PP^{-1} = I.$$

Let $Q = P^{-1}$ so that $Q^{-1} = P$. Then $Q^{-1}BQ = A$.

$$Q = P^{-1} = \begin{bmatrix} \frac{1}{2} & \frac{1}{2} \\ \frac{1}{2} & -\frac{1}{2} \end{bmatrix}.$$

Thus,

$$Q^{-1} = \frac{1}{\det(Q)} \text{adj}(Q)$$

$$Q^{-1} = \frac{1}{-\frac{1}{2}} \begin{bmatrix} -\frac{1}{2} & -\frac{1}{2} \\ -\frac{1}{2} & \frac{1}{2} \end{bmatrix} = \begin{bmatrix} 1 & 1 \\ 1 & -1 \end{bmatrix}$$

$$Q^{-1}BQ = \begin{bmatrix} 1 & 1 \\ 1 & -1 \end{bmatrix} \begin{bmatrix} \frac{3}{2} & -\frac{1}{2} \\ \frac{1}{2} & \frac{1}{2} \end{bmatrix} \begin{bmatrix} \frac{1}{2} & \frac{1}{2} \\ \frac{1}{2} & -\frac{1}{2} \end{bmatrix}$$

$$= \begin{bmatrix} 1 & 1 \\ 1 & -1 \end{bmatrix} \begin{bmatrix} \frac{1}{2} & 1 \\ \frac{1}{2} & 0 \end{bmatrix}$$

$$= \begin{bmatrix} 1 & 1 \\ 0 & 1 \end{bmatrix}$$

$$= A$$

Thus, if B is similar to A, then A is similar to B.

(d) Let $K = 2$; check that $P^{-1}A^2P = B^2$.

Then

$$B^2 = \begin{bmatrix} \frac{3}{2} & -\frac{1}{2} \\ \frac{1}{2} & \frac{1}{2} \end{bmatrix} \begin{bmatrix} \frac{3}{2} & -\frac{1}{2} \\ \frac{1}{2} & \frac{1}{2} \end{bmatrix}$$

$$= \begin{bmatrix} 2 & -1 \\ 1 & 0 \end{bmatrix}$$

$$A^2 = \begin{bmatrix} 1 & 1 \\ 0 & 1 \end{bmatrix} \begin{bmatrix} 1 & 1 \\ 0 & 1 \end{bmatrix}$$

$$= \begin{bmatrix} 1 & 2 \\ 0 & 1 \end{bmatrix}$$

Then,

$$P^{-1}A^2P = \begin{bmatrix} \frac{1}{2} & \frac{1}{2} \\ \frac{1}{2} & -\frac{1}{2} \end{bmatrix} \begin{bmatrix} 1 & 2 \\ 0 & 1 \end{bmatrix} \begin{bmatrix} 1 & 1 \\ 1 & -1 \end{bmatrix}$$

$$= \begin{bmatrix} \frac{1}{2} & \frac{1}{2} \\ \frac{1}{2} & -\frac{1}{2} \end{bmatrix} \begin{bmatrix} 3 & -1 \\ 1 & -1 \end{bmatrix}$$

$$= \begin{bmatrix} 2 & -1 \\ 1 & 0 \end{bmatrix}$$

$$= B^2$$

Suppose $K = 3$:

$$A^3 = A^2 A = \begin{bmatrix} 1 & 2 \\ 0 & 1 \end{bmatrix}\begin{bmatrix} 1 & 1 \\ 0 & 1 \end{bmatrix}$$

$$= \begin{bmatrix} 1 & 3 \\ 0 & 1 \end{bmatrix}$$

$$B^3 = B^2 B = \begin{bmatrix} 2 & -1 \\ 1 & 0 \end{bmatrix}\begin{bmatrix} \frac{3}{2} & -\frac{1}{2} \\ \frac{1}{2} & \frac{1}{2} \end{bmatrix}$$

$$= \begin{bmatrix} \frac{5}{2} & -\frac{3}{2} \\ \frac{3}{2} & -\frac{1}{2} \end{bmatrix}$$

Then,

$$P^{-1}A^3 P = \begin{bmatrix} \frac{1}{2} & \frac{1}{2} \\ \frac{1}{2} & -\frac{1}{2} \end{bmatrix}\begin{bmatrix} 1 & 3 \\ 0 & 1 \end{bmatrix}\begin{bmatrix} 1 & 1 \\ 1 & -1 \end{bmatrix}$$

$$= \begin{bmatrix} \frac{1}{2} & \frac{1}{2} \\ \frac{1}{2} & -\frac{1}{2} \end{bmatrix}\begin{bmatrix} 4 & -2 \\ 1 & -1 \end{bmatrix}$$

$$= \begin{bmatrix} \frac{5}{2} & -\frac{3}{2} \\ \frac{3}{2} & -\frac{1}{2} \end{bmatrix}$$

$$= B^3$$

In general for any positive integer k, $B^k = P^{-1}A^k P$ if $B = P^{-1}AP$. To prove this rigorously for any matrices A and B and an invertible matrix P, use an inductive argument. Given $P^{-1}AP = B$, show

$P^{-1}A^kP = B^k$. Take $n = 1$; $P^{-1}AP = B$. When $n = 2$; $P^{-1}AP = B$ gives $B^2 = P^{-1}APP^{-1}$ so $B^2 = P^{-1}A^2P$ since $PP^{-1} = I$.

Assume $P^{-1}A^kP = B^k$ is true for $k = n$; show that it is true for $k = n + 1$. $P^{-1}A^nP = B^n$ so, since $B^{n+1} = B^n \cdot B = P^{-1}A^nPB = P^{-1}A^nPP^{-1}AP = P^{-1}A^{n+1}P$, $B^{n+1} = P^{-1}A^{n+1}P$.

From this it follows that if B is similar to A, then B^k is similar to A^k. Observe that the powers of A are easy to find. Direct calculation gives:

$$A^3 \begin{bmatrix} 1 & 3 \\ 0 & 1 \end{bmatrix}; \quad A^4 = \begin{bmatrix} 1 & 4 \\ 0 & 1 \end{bmatrix}.$$

In general, we obtain the formula

$$A^k = \begin{bmatrix} 1 & k \\ 0 & 1 \end{bmatrix}.$$

Again, to be rigorous, one would need to use an inductive argument. To find B^k, use the formula $B = P^{-1}A^kB$. Thus,

$$
\begin{aligned}
B^k &= P^{-1} \begin{bmatrix} 1 & k \\ 0 & 1 \end{bmatrix} \begin{bmatrix} 1 & 1 \\ 1 & -1 \end{bmatrix} \\
&= P^{-1} \begin{bmatrix} 1+k & 1-k \\ 1 & -1 \end{bmatrix} \\
&= \begin{bmatrix} \frac{1}{2} & \frac{1}{2} \\ \frac{1}{2} & -\frac{1}{2} \end{bmatrix} \begin{bmatrix} 1+k & 1-k \\ 1 & -1 \end{bmatrix} \\
&= \begin{bmatrix} 1+\frac{k}{2} & -\frac{k}{2} \\ \frac{k}{2} & 1-\frac{k}{2} \end{bmatrix}
\end{aligned}
$$

4.5.1 What the Determinant Measures

The absolute value of the determinant of a 2×2 matrix measures how much area is distorted by the linear tranformation represented by the matrix.

EXAMPLE

If T is represented by $\begin{bmatrix} 1 & 2 \\ 0 & 2 \end{bmatrix}$, then

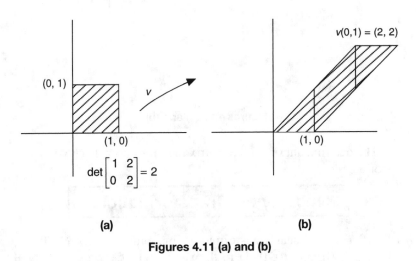

$$\det \begin{bmatrix} 1 & 2 \\ 0 & 2 \end{bmatrix} = 2$$

(a)   (b)

Figures 4.11 (a) and (b)

Under the transformation T, the area of the unit square has gone from 1 to 2.1.

This gives a visual demonstration of why the determinant of a matrix with two rows which are multiples of each other is zero. In this case, the unit box is mapped to a line which has zero area.

EXAMPLE

Let $A = [2'\ 2']$

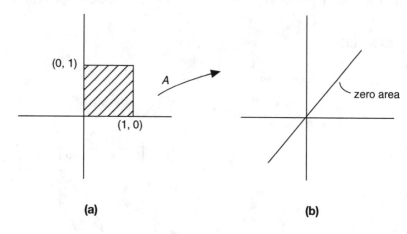

(a)   (b)

Figures 4.12 (a) and (b)

The determinant of a 3×3 matrix measures how much volume is distorted.

Quiz: Linear Transformations

1. If T is a linear transformation mapping vectors $(1, 0, 0)$, $(0, 1, 0)$, and $(0, 0, 1)$ to the vectors $(1, 2, 3)$, $(2, 3, 1)$, and $(1, 1, -2)$, respectively, which vector is the image of the vector $(3, -2, 1)$ under T?

    (A) $(1, 1, 7)$ (D) $(0, 1, 9)$

    (B) $(1, 0, 5)$ (E) $(1, 7, 0)$

    (C) $(0, 1, 5)$

2. The kernel of the linear transformation

    $$T(x, x, x) = (2x - x, x - x)$$

    is

(A) $(2x, x, x)$.   (D) $(x, x, 2x)$.

(B) (x, x, x).   (E) $\left(\dfrac{x}{2}, x, x\right)$.

(C) $\left(x, x, \dfrac{x}{2}\right)$.

3. The product $A\,B$ of the two matrices

$$A = \begin{bmatrix} 1 & 2 \\ -1 & 3 \end{bmatrix} \text{ and } B = \begin{bmatrix} 0 & 2 \\ 2 & -1 \end{bmatrix}$$

defines a linear transformation that carries the point (x, y) into the point with coordinates

(A) $(x - y, 2x + 3y)$.   (D) $(4x, 6x + 5y)$.

(B) $(4x, 6x - 5y)$.   (E) $(6x - 5y, 4x)$.

(C) $(2y, 2x - y)$.

4. Let T represent a nonsingular linear transformation from E into E. Which of the following is *not* true?

(A) Null space of $T = \{0\}$.

(B) T is one-to-one.

(C) Dimension of null space is zero: $dim\,(N(T)) = 0$.

(D) Dimension of range space is n: $dim\,(R(T)) = n$.

(E) $dim\,(N(T)) = dim\,(R(T))$.

5. Let $T : R \to R$ be defined by

$$T(x, y) = \begin{bmatrix} 2x - y \\ x + 3y \end{bmatrix}$$

Find the adjoint T^* of T.

(A) $\begin{bmatrix} 2x + y \\ -x + 3y \end{bmatrix}$.

(D) $\begin{bmatrix} \dfrac{x}{2} - y \\ -x + \dfrac{y}{3} \end{bmatrix}$.

(B) $\begin{bmatrix} x + 2y \\ x - 3y \end{bmatrix}$.

(E) $\begin{bmatrix} 3x - y \\ x + 2y \end{bmatrix}$.

(C) $\begin{bmatrix} 2x + y \\ x - 3y \end{bmatrix}$.

6. We define a bilinear form in the following manner: Let V be a finite dimensional vector space over a field F. A bilinear form $b(v; \bar{v})$ is a function $b: V \times V \to F$ which satisfies

1) $b(\alpha v_1 + \beta v_2; \bar{v}) = \alpha b(v_1; \bar{v}) + \beta b(v_2; \bar{v})$

2) $b(v; \delta \bar{v}_1 \, v \bar{v}_2) = \delta(v; \bar{v}_1) + v \, b(v; \bar{v}_2)$

where $v_1, v_2, \bar{v}_1, \bar{v}_2 \in V$ and $\alpha, \beta, v, \in F$. The matrix $B = (b)$ is given by $b = b(u, u)$ for $1 \le i \le 2; \ 1 \le j \le 2$.

Let $b : R \times R \to$ be the bilinear form defined by

$$b(X; Y) = x_1 y_1 - 2x_1 y_2 + x_2 y_1 + 3x_2 y_2$$

where $X = (x, x)$ and $Y = (y, y)$. Find the 2×2 matrix B of T relative to the basis $U = \{u, u\}$ where $u = (0, 1)$ and $u = (1, 1)$.

(A) $\begin{bmatrix} 5 & -3 \\ 0 & 2 \end{bmatrix}$.

(D) $\begin{bmatrix} 0 & 4 \\ -1 & 3 \end{bmatrix}$.

(B) $\begin{bmatrix} 2 & 2 \\ -1 & 1 \end{bmatrix}$.

(E) $\begin{bmatrix} 3 & -1 \\ 1 & 2 \end{bmatrix}$.

(C) $\begin{bmatrix} -1 & 4 \\ 2 & 3 \end{bmatrix}$.

7. Let T be a linear transformation from R to R. If $\vec{u}$ and $\vec{v}$ are two orthogonal vectors in R, which of the following pairs of vectors must be orthogonal to one another?

    (A) $r\vec{u}$ and $s\vec{v}$ for all real r and s.

    (B) $\vec{u} + \vec{v}$ and $\vec{u} - \vec{v}$.

    (C) $T\vec{u}$ and $\vec{u}$.

    (D) $T\vec{v}$ and $\vec{v}$.

    (E) $T\vec{u}$ and $T\vec{v}$.

8. Let T be a linear transformation from a vector space V of dimension 11 *onto* a vector space W of dimension 7. What is the dimension of the null space of T?

    (A) 0.   (D) 4.

    (B) 2.   (E) 5.

    (C) 3.

9. Which of the following matrices is orthogonal?

    (A) $\begin{bmatrix} 0 & 1 \\ -1 & 0 \end{bmatrix}$.   (D) $\begin{bmatrix} 2 & 1 \\ 0 & 1 \end{bmatrix}$.

    (B) $\begin{bmatrix} 1 & 2 \\ 0 & 1 \end{bmatrix}$.   (E) $\begin{bmatrix} 1 & -1 \\ 1 & -1 \end{bmatrix}$.

    (C) $\begin{bmatrix} 1 & 0 \\ 1 & 0 \end{bmatrix}$.

10. The linear transformation of the unit basis vectors of V produces the vectors $(-2, 0, 0)$, $(0, 4, 0)$, and $(0, 0, -1)$. Under this transformation, the vector $(2, 3, 1)$ is carried into the vector

(A) $(1, 1, 1)$.

(D) $(-4, 12, 1)$.

(B) $(2, 3, 1)$.

(E) $(4, -12, 1)$.

(C) $(-2, 4, -1)$.

ANSWER KEY

1.	(C)	6.	(D)
2.	(D)	7.	(A)
3.	(B)	8.	(D)
4.	(E)	9.	(A)
5.	(A)	10.	(D)

CHAPTER 5

Eigenvalues and Eigenvectors

5.1 Definitions of Eigenvalues and Eigenvectors

If A is an $n \times n$ matrix, then the non-zero vector $\vec{X} \in R^n$ is called an eigenvector of A if $A\vec{x} = \lambda \vec{x}$, where λ is a scalar called the eigenvalue of A.

If A is a square matrix, then the characteristic equation of A is defined to be $\det(\lambda I - A) = 0$. When expanded, the $\det(\lambda I - A)$ is called the characteristic polynomial of A.

Problem Solving Examples:

Find the characteristic polynomials and the eigenvalues of the matrices.

a) $A = \begin{bmatrix} 2 & 3 \\ 1 & 4 \end{bmatrix}$; b) $B = \begin{bmatrix} \cos\alpha & \sin\alpha \\ -\sin\alpha & \cos\alpha \end{bmatrix}$; c) $C = \begin{bmatrix} 1 & 2 & 3 \\ 2 & 1 & 3 \\ 3 & 3 & 6 \end{bmatrix}$

a) The characteristic polynomial of A is

$$\det(\lambda I - A) = \det \begin{bmatrix} \lambda - 2 & -3 \\ -1 & \lambda - 4 \end{bmatrix}$$
$$= (\lambda - 2)(\lambda - 4) - 3$$
$$= \lambda^2 - 6\lambda + 5$$
$$= (\lambda - 1)(\lambda - 5)$$

The characteristic equation is $(\lambda - 1)(\lambda - 5) = 0$. Then the characteristic values are $\lambda = 1$ and $\lambda = 5$.

The zeros of the characteristic polynomial of a matrix are also called characteristic numbers or, proper values.

b) The characteristic polynomial of B is:

$$\det(\lambda I - B) = \det \begin{bmatrix} \lambda - \cos\alpha & -\sin\alpha \\ \sin\alpha & \lambda - \cos\alpha \end{bmatrix}$$
$$= (\lambda - \cos\alpha)(\lambda - \cos\alpha) + \sin^2\alpha$$
$$= \lambda^2 - 2\lambda\cos\alpha + \cos^2\alpha + \sin^2\alpha$$

But $\sin^2\alpha + \cos^2\alpha = 1$. Therefore, $\det(\lambda I - B) = \lambda^2 - 2\lambda\cos\alpha + 1$. The characteristic equation is

$$\lambda^2 - 2\cos\alpha \ \lambda + 1 = 0.$$

We know the root of an equation, $ax^2 + bx + c = 0$, is

$$x = \frac{-b \pm \sqrt{b^2 - 4ac}}{2a}$$

Thus,

$$\lambda = \frac{2\cos\alpha \pm \sqrt{4\cos^2\alpha - 4}}{2}$$

or

$$\lambda = \frac{2\cos\alpha \pm 2\sqrt{\cos^2\alpha - 1}}{2}$$
$$= \cos\alpha \pm \sqrt{\cos^2\alpha - 1}$$

But, $\cos^2 \alpha - 1 = -\sin^2 \alpha$; therefore, $\lambda = \cos \alpha \pm \sqrt{-\sin^2 \alpha}$ or $\lambda = \cos \alpha \pm i \sin \alpha$.

c)   The characteristic polynomial of C is:

$$\det(\lambda I - C) = \det \begin{bmatrix} \lambda - 1 & -2 & -3 \\ -2 & \lambda - 1 & -3 \\ -3 & -3 & \lambda - 6 \end{bmatrix}$$

$$= (\lambda - 1)\begin{vmatrix} \lambda - 1 & -3 \\ -3 & \lambda - 6 \end{vmatrix} - (-2)\begin{vmatrix} -2 & -3 \\ -3 & \lambda - 6 \end{vmatrix} + (-3)\begin{vmatrix} -2 & \lambda - 1 \\ -3 & -3 \end{vmatrix}$$

$$= (\lambda - 1)\left[(\lambda - 1)(\lambda - 6) - 9\right] + 2\left[-2(\lambda - 6) - 9\right] - 3\left[6 + 3(\lambda - 1)\right]$$

$$= (\lambda - 1)\left[\lambda^2 - 7\lambda - 3\right] + 2\left[-2\lambda + 3\right] - 3\left[3 + 3\lambda\right]$$

$$= \lambda^3 - 8\lambda^2 + 4\lambda + 3 - 4\lambda + 6 - 9 - 9\lambda$$

$$= \lambda^3 - 8\lambda^2 - 9\lambda$$

$$= \lambda(\lambda - 9)(\lambda + 1)$$

The characteristic equation is $\lambda(\lambda - 9)(\lambda + 1) = 0$. Then, the characteristic values are $\lambda = 0$, $\lambda = -1$, and $\lambda = 9$.

 Find the real eigenvalues of the matrix

$$A = \begin{bmatrix} -2 & -1 \\ 5 & 2 \end{bmatrix}.$$

 Form the matrix

$$\lambda I - A = \lambda \begin{bmatrix} 1 & 0 \\ 0 & 1 \end{bmatrix} - \begin{bmatrix} -2 & -1 \\ 5 & 2 \end{bmatrix}$$

$$= \begin{bmatrix} \lambda + 2 & 1 \\ -5 & \lambda - 2 \end{bmatrix}$$

Take its determinant to obtain the characteristic polynomial of A:

$$f(\lambda) = \det(\lambda I - A)$$
$$= \det \begin{vmatrix} \lambda + 2 & 1 \\ -5 & \lambda - 2 \end{vmatrix}$$
$$= (\lambda + 2)(\lambda - 2) + 5$$
$$= \lambda^2 + 1$$

The eigenvalues of A must, therefore, satisfy the quadratic equation $\lambda^2 + 1 = 0$. Since the only solutions to this equation are the imaginary numbers $\lambda = i$ and $\lambda = -i$, A has no real eigenvalues.

Find the real eigenvalues of A and their associated eigenvectors when

$$A = \begin{bmatrix} 1 & 1 \\ -2 & 4 \end{bmatrix}.$$

We wish to find all real numbers λ and all non-zero vectors $\vec{x} = \begin{bmatrix} x_1 \\ x_2 \end{bmatrix}$ such that $A\vec{x} = \lambda\vec{x}$:

$$\begin{bmatrix} 1 & 1 \\ -2 & 4 \end{bmatrix}\begin{bmatrix} x_1 \\ x_2 \end{bmatrix} = \lambda\begin{bmatrix} x_1 \\ x_2 \end{bmatrix}$$

The above matrix equation is equivalent to the homogeneous system,

$$x_1 + x_2 = \lambda x_1$$
$$-2x_1 + 4x_2 = \lambda x_2$$

or

$$(\lambda - 1)x_1 - x_2 = 0 \tag{1}$$
$$2x_1 + (\lambda - 4)x_2 = 0$$

Recall that a homogeneous system has a non-zero solution if and only if the determinant of the matrix of coefficients is zero. Thus,

$$\begin{vmatrix} \lambda - 1 & 1 \\ 2 & \lambda - 4 \end{vmatrix} = 0$$

or

$$(\lambda - 1)(\lambda - 4) + 2 = 0.$$

Therefore,

$$\lambda^2 - 5\lambda + 6 = 0$$

or

$$(\lambda - 3)(\lambda - 2) = 0.$$

Hence, $\lambda_1 = 2$ and $\lambda_2 = 3$ are the eigenvalues of A. To find an eigenvector of A associated with $\lambda_1 = 2$, form the linear system:

$$A\vec{x} = 2\vec{x}$$

or

$$\begin{bmatrix} 1 & 1 \\ -2 & 4 \end{bmatrix} \begin{bmatrix} x_1 \\ x_2 \end{bmatrix} = 2 \begin{bmatrix} x_1 \\ x_2 \end{bmatrix}$$

This gives

$$x_1 + x_2 = 2x_1 \qquad \text{or} \qquad (2-1)x_1 - x_2 = 0$$
$$-2x_1 + 4x_2 = 2x_2 \qquad \qquad 2x_1 + (2-4)x_2 = 0$$

or

$$x_1 - x_2 = 0$$
$$2x_1 - 2x_2 = 0 \qquad \text{or, simply, } x_1 - x_2 = 0.$$

Observe that we could have obtained this last linear system by substituting $\lambda = 2$ in (1). It can be seen that any vector in R^2 of the form

$$x = k \begin{bmatrix} 1 \\ 1 \end{bmatrix},$$

k a scalar, is an eigenvector of A associated with $\lambda_1 = 2$. Thus,

$$x_1 = \begin{bmatrix} 1 \\ 1 \end{bmatrix}$$

is an eigenvector of A associated with $\lambda_1 = 2$. Similarly, for $\lambda_2 = 3$, we obtain from (1):

$$(3-1)x_1 - x_2 = 0 \qquad \text{or} \qquad 2x_1 = x_2 = 0$$
$$2x_1 + (3-4)x_2 = 0 \qquad\qquad 2x_1 - x_2 = 0$$

Thus,

$$x_2 = \begin{bmatrix} 1 \\ 2 \end{bmatrix}$$

is an eigenvector of A associated with $\lambda_2 = 3$.

Q Find the eigenvalues and the corresponding eigenvectors of A where

$$A = \begin{bmatrix} 0 & \frac{1}{2} \\ \frac{1}{2} & 0 \end{bmatrix}.$$

A An eigenvalue of A is a scalar λ such that $A\vec{x} = \lambda\vec{x}$ for some non-zero vectors $\vec{x}$. This may be converted to $(\lambda I - A)\vec{x} = 0$ which implies $\det(\lambda I - A) = 0$, the characteristic equation. The roots of this equation yield the required eigenvalues.

$$\lambda I - A = \begin{bmatrix} \lambda & 0 \\ 0 & \lambda \end{bmatrix} - \begin{bmatrix} 0 & \frac{1}{2} \\ \frac{1}{2} & 0 \end{bmatrix} = \begin{bmatrix} \lambda & -\frac{1}{2} \\ -\frac{1}{2} & \lambda \end{bmatrix}$$

$$\det(\lambda I - A) = \lambda^2 - \frac{1}{4}.$$

Then, the characteristic equation is $\lambda^2 - \frac{1}{4} = 0$ and the eigenvalues are $\lambda_1 = \frac{1}{2}$ and $\lambda_2 = \frac{1}{2}$. Substitute $\lambda = \frac{-1}{2}$ in the equation $(\lambda I - A)\vec{x} = 0$ to obtain the corresponding eigenvectors. $(\frac{1}{2}I - A)\vec{x} = 0$

$$\begin{bmatrix} \frac{1}{2} & -\frac{1}{2} \\ -\frac{1}{2} & \frac{1}{2} \end{bmatrix} \begin{bmatrix} x_1 \\ x_2 \end{bmatrix} = \begin{bmatrix} 0 \\ 0 \end{bmatrix}$$

or

$$\frac{1}{2}x_1 - \frac{1}{2}x_2 = 0 \qquad \text{or} \qquad x_1 - x_2 = 0$$
$$-\frac{1}{2}x_1 + \frac{1}{2}x_2 = 0$$

Thus,

$$x_1 = \begin{bmatrix} 1 \\ 1 \end{bmatrix}$$ is an eigenvector of A associated with the eigenvalue $\lambda_1 = \frac{1}{2}$. Now, let $\lambda = -\frac{1}{2}$. Then,

$$\begin{bmatrix} -\frac{1}{2} & -\frac{1}{2} \\ -\frac{1}{2} & -\frac{1}{2} \end{bmatrix} \begin{bmatrix} x_1 \\ x_2 \end{bmatrix} = \begin{bmatrix} 0 \\ 0 \end{bmatrix}.$$

Therefore,

$$-\frac{1}{2}x_1 - \frac{1}{2}x_2 = 0; \qquad \text{or} \qquad x_1 + x_2 = 0$$
$$-\frac{1}{2}x_1 - \frac{1}{2}x_2 = 0$$

Then,

$$x_2 = \begin{bmatrix} 1 \\ -1 \end{bmatrix}$$ is an eigenvector of A associated with the eigenvalue $\lambda_2 = -\frac{1}{2}$. If we let $T: R^2 \to R^2$ be defined by

$$T(\vec{x}) = A\vec{x} = \begin{bmatrix} 0 & \frac{1}{2} \\ \frac{1}{2} & 0 \end{bmatrix} \begin{bmatrix} x_1 \\ x_2 \end{bmatrix}$$

then Figure 5.1 shows that $\vec{x_1}$ and $T\left(\vec{x_1}\right)$ are parallel and that $\vec{x_2}$ and $T\left(\vec{x_2}\right)$ are parallel also. This illustrates the fact that if $\vec{x}$ is an eigenvector of A, then $\vec{x}$ and $A\vec{x}$ are parallel.

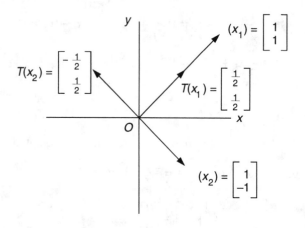

Figure 5.1

Identify the following quadratic forms by finding their eigenvalues.

1) $Q(x, y) = 2x^2 - 4xy - y^2$

2) $Q(x, y) = 9x^2 + 6xy + y^2$

Every quadratic form has associated with it a symmetric matrix S. The coordinate change that reduces the quadric surface to a recognizable conic is exactly the change of basis which diagonalizes the matrix S. It is the diagonalized matrix whose elements are eigenvalues of S that represents the reduced quadratic form. The type of conic surface represented is determined by the nature of the eigenvalues. Thus, it makes sense to say that we can "identify" a quadratic form by its eigenvalues.

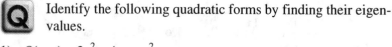

1) $Q(x, y) = 2x^2 - 4xy - y^2 = \begin{bmatrix} x & y \end{bmatrix}\begin{bmatrix} 2 & -2 \\ -2 & -1 \end{bmatrix}\begin{bmatrix} x \\ y \end{bmatrix}$ \hfill (1)

To find the eigenvalues of the matrix in (1), set

$$\begin{bmatrix} 2 & -2 \\ -2 & -1 \end{bmatrix} \begin{bmatrix} x \\ y \end{bmatrix} = \lambda \begin{bmatrix} x \\ y \end{bmatrix}$$

or

$$\begin{bmatrix} 2-\lambda & 2 \\ -2 & -1-\lambda \end{bmatrix} \begin{bmatrix} x \\ y \end{bmatrix} = \begin{bmatrix} 0 \\ 0 \end{bmatrix}. \tag{2}$$

System (2) can have non-trivial solutions only if:

$$\det \begin{bmatrix} 2-\lambda & -2 \\ -2 & -1-\lambda \end{bmatrix} = 0$$

or $\qquad (\lambda - 3)(\lambda + 2) = 0.$

Hence, the eigenvalues are $\lambda = 3$ and $\lambda = -2$, and there exists, with respect to some basis, a diagonal matrix:

$$D = \begin{bmatrix} 3 & 0 \\ 0 & -2 \end{bmatrix} \tag{3}$$

which represents the quadratic form. To show this, first find two orthogonal eigenvectors associated with eigenvalues. If $\lambda = 3$, (2) becomes the system:

$$\begin{bmatrix} -1 & -2 \\ -2 & -4 \end{bmatrix} \begin{bmatrix} x \\ y \end{bmatrix} = \begin{bmatrix} 0 \\ 0 \end{bmatrix}.$$

By solving the system, we find that an eigenvector associated with $\lambda = 3$ is $[x, y] = [2, -1]$. If $\lambda = -2$, by substitution (2) becomes

$$\begin{bmatrix} 4 & -2 \\ -2 & 1 \end{bmatrix} \begin{bmatrix} x \\ y \end{bmatrix} = \begin{bmatrix} 0 \\ 0 \end{bmatrix},$$

and solving, we find an eigenvector $[x, y] = [1, 2]$. These two eigenvectors are linearly independent. In fact, they are orthogonal since $< [1, 2]), ([2, -1] > = 0$. If normalized, they will form an orthonormal basis, since both vectors are of norm $\sqrt{5}$.

$$B = \left\{ \frac{1}{\sqrt{5}}[2, -1], \frac{1}{\sqrt{5}}[1, 2] \right\}$$

is an orthonormal basis. With respect to this basis, the quadratic form has the diagonal matrix

$$\begin{bmatrix} 3 & 0 \\ 0 & -2 \end{bmatrix}$$

so,

$$\begin{bmatrix} 3 & 0 \\ 0 & -2 \end{bmatrix} \begin{bmatrix} x^2 \\ y^2 \end{bmatrix} = 3x^2 - 2y^2$$

which is the equation for a hyperbola.

2) Note that in 1), the diagonal matrix had the eigenvalues of the matrix in (1) as its diagonal entries:

$$Q(x, y) = 9x^2 + 6xy + y^2 = \begin{bmatrix} x & y \end{bmatrix} \begin{bmatrix} 9 & 3 \\ 3 & 1 \end{bmatrix} \begin{bmatrix} x \\ y \end{bmatrix}. \qquad (4)$$

The eigenvalues of the matrix in (4) are found by solving

$$\begin{bmatrix} 9 & 3 \\ 3 & 1 \end{bmatrix} \begin{bmatrix} x \\ y \end{bmatrix} = \lambda \begin{bmatrix} x \\ y \end{bmatrix}$$

which yields

$$\begin{bmatrix} 9 - \lambda & 3 \\ 3 & 1 - \lambda \end{bmatrix} \begin{bmatrix} x \\ y \end{bmatrix} = \begin{bmatrix} 0 \\ 0 \end{bmatrix}.$$

Hence, $\lambda(\lambda - 10) = 0$ or, $\lambda = 0$ and $\lambda = 10$. Thus, the diagonal matrix representing the quadratic form is

$$D = \begin{bmatrix} 10 & 0 \\ 0 & 0 \end{bmatrix},$$

and the equation of the conic is $10x^2$. This is the equation of a parabola.

We could also say that if there are two eigenvalues which are both strictly positive, the surface is an ellipse.

5.2   Diagonalization

If A is a square matrix and there exists a matrix P such that $P^{-1}AP$ is diagonal, then A is diagonalizable and P diagonalizes A.

THEOREM

If A is an $n \times n$ matrix and A is diagonalizable, then A has n linearly independent eigenvectors.

The following is the procedure for diagonalizing an $n \times n$ matrix A:

a) Find the set of n linearly independent eigenvectors of A, $\left\{ \overrightarrow{v_1}, \overrightarrow{v_2}, \ldots, \overrightarrow{v_n} \right\}$.

b) Form the matrix P having $\overrightarrow{v_1}, \overrightarrow{v_2}, \ldots, \overrightarrow{v_n}$ as its column vectors.

c) The matrix $P^{-1}AP$ will then be diagonal, with $\lambda_1, \lambda_2, \ldots, \lambda_n$ as its diagonal entries where λ_i is the eigenvalue corresponding to $\overrightarrow{v_i}, i = 1, 2, \ldots, n$.

THEOREM

If $\overrightarrow{v_1}, \overrightarrow{v_2}, \ldots, \overrightarrow{v_n}$ are eigenvectors corresponding to distinct eigenvalues $\lambda_1, \lambda_2, \ldots, \lambda_n$, then $\left\{ \overrightarrow{v_1}, \overrightarrow{v_2}, \ldots, \overrightarrow{v_n} \right\}$ is a linearly independent set.

THEOREM

If an $n \times n$ matrix has n distinct eigenvalues, then it is diagonalizable.

Problem Solving Examples:

Given that

$$A = \begin{bmatrix} 1 & 1 \\ -2 & 4 \end{bmatrix},$$

find an invertible matrix P such that $P^{-1}AP$ is a diagonal matrix D.

The characteristic equation of the matrix A is $\det(\lambda I - A) = 0$. Thus,

$$\det\begin{bmatrix} \lambda - 1 & -1 \\ 2 & \lambda - 4 \end{bmatrix} = 0,$$

or

$$(\lambda - 1)(\lambda - 4) + 2 = 0$$
$$\lambda^2 - 5\lambda + 6 = 0$$

or

$$(\lambda - 3)(\lambda - 2) = 0.$$

The eigenvalues are therefore $\lambda_1 = 2$ and $\lambda_2 = 3$. To obtain the eigenvector corresponding to the eigenvalue $\lambda_1 = 2$, solve the equation $(2I - A)\vec{x} = 0$ for $\vec{x}$:

$$\begin{bmatrix} 1 & -1 \\ 2 & -2 \end{bmatrix}\begin{bmatrix} x_1 \\ x_2 \end{bmatrix} = \begin{bmatrix} 0 \\ 0 \end{bmatrix},$$

or

$$\begin{array}{lll} x_1 - x_2 = 0 & \text{which gives} & x_1 - x_2 = 0 \text{ or} \\ 2x_1 - 2x_2 = 0 & & x_1 = x_2 \end{array}$$

Thus, $\vec{x}_1 = \begin{bmatrix} 1 \\ 1 \end{bmatrix}$ is an eigenvector of 2. Similarly, an eigenvector $\vec{x}_2$

corresponding to the eigenvalue $\lambda = 3$ is $\begin{bmatrix} 1 \\ 2 \end{bmatrix}$. Since the eigenvectors

$\vec{x}_1$ and $\vec{x}_2$ are linearly independent, A is diagonalizable. It is possible to see that $\vec{x}_1$ and $\vec{x}_2$ are independent because $a_1\vec{x}_1 + a_2\vec{x}_2 = 0$ has only the trivial solution, $a_1 = a_2 = 0$. Let P be the matrix whose columns are $\vec{x}_1$ and $\vec{x}_2$:

$$P = \begin{bmatrix} 1 & 1 \\ 1 & 2 \end{bmatrix}$$

and

$$P^{-1} = \begin{bmatrix} 2 & -1 \\ -1 & 1 \end{bmatrix}$$

(using the adjoint method). Thus,

$$P^{-1}AP = \begin{bmatrix} 2 & -1 \\ -1 & 1 \end{bmatrix}\begin{bmatrix} 1 & 1 \\ -2 & 4 \end{bmatrix}\begin{bmatrix} 1 & 1 \\ 1 & 4 \end{bmatrix} = \begin{bmatrix} 2 & 0 \\ 0 & 3 \end{bmatrix}.$$

On the other hand, if we let $\lambda_1 = 3$ and $\lambda_2 = 2$, then $x_1 = \begin{bmatrix} 1 \\ 2 \end{bmatrix}$ and $x_2 = \begin{bmatrix} 1 \\ 1 \end{bmatrix}$. In that case,

$$P = \begin{bmatrix} 1 & 1 \\ 2 & 1 \end{bmatrix} \text{ and } P^{-1} = \begin{bmatrix} -1 & 1 \\ 2 & -1 \end{bmatrix},$$

and

$$P^{-1}AP = \begin{bmatrix} -1 & 1 \\ 2 & -1 \end{bmatrix}\begin{bmatrix} 1 & 1 \\ -2 & 4 \end{bmatrix}\begin{bmatrix} 1 & 1 \\ 2 & 1 \end{bmatrix} = \begin{bmatrix} 3 & 0 \\ 0 & 2 \end{bmatrix}.$$

 Find a matrix P that diagonalizes

$$A = \begin{bmatrix} 3 & -2 & 0 \\ -2 & 3 & 0 \\ 0 & 0 & 5 \end{bmatrix}.$$

 The characteristic equation of A is

$$\det \begin{bmatrix} \lambda - 3 & 2 & 0 \\ 2 & \lambda - 3 & 0 \\ 0 & 0 & \lambda - 5 \end{bmatrix} = 0.$$

Expanding along the third row,

$$(\lambda - 5)\begin{vmatrix} \lambda - 3 & 2 \\ 2 & \lambda - 3 \end{vmatrix} = 0,$$

or

$$(\lambda - 5)\left[(\lambda - 3)^2 - 4\right] = 0,$$

or

$$(\lambda - 5)(\lambda - 1)(\lambda - 5) = 0.$$

The eigenvalues of A are $\lambda_1 = 1$ and $\lambda_2 = 5$.

Solve the equation $(5I - A)\vec{y} = 0$ to obtain the eigenvectors corresponding to $\lambda = 5$:

$$\begin{bmatrix} 2 & 2 & 0 \\ 2 & 2 & 0 \\ 0 & 0 & 0 \end{bmatrix}\begin{bmatrix} y_1 \\ y_2 \\ y_3 \end{bmatrix} = \begin{bmatrix} 0 \\ 0 \\ 0 \end{bmatrix}.$$

Solving the above system yields:

$$y_1 = -s, y_2 = s, y_3 = t,$$

or

$$y = \begin{bmatrix} -s \\ s \\ t \end{bmatrix} = \begin{bmatrix} -s \\ s \\ 0 \end{bmatrix} + \begin{bmatrix} 0 \\ 0 \\ t \end{bmatrix} = s\begin{bmatrix} -1 \\ 1 \\ 0 \end{bmatrix} + t\begin{bmatrix} 0 \\ 0 \\ 1 \end{bmatrix}.$$

Thus,

$$x_1 = \begin{bmatrix} -1 \\ 1 \\ 0 \end{bmatrix} \text{ and } x_2 = \begin{bmatrix} 0 \\ 0 \\ 1 \end{bmatrix}$$ are two linearly independent

eigenvectors corresponding to $\lambda = 5$.

Similarly, we find that $\vec{x}_3 = \begin{bmatrix} 1 \\ 1 \\ 0 \end{bmatrix}$ is an eigenvector associated with

$\lambda - 1$. Thus, $\vec{x}_1$, $\vec{x}_2$, and $\vec{x}_3$ are linearly independent vectors such that

$$P = \begin{bmatrix} -1 & 0 & 1 \\ 1 & 0 & 1 \\ 0 & 1 & 0 \end{bmatrix}.$$

To find P^{-1}, use the adjoint method:

$$P^{-1} = \frac{1}{\det(P)} \text{ adj } (P).$$

We find

$$P^{-1} = \begin{bmatrix} -\frac{1}{2} & \frac{1}{2} & 0 \\ 0 & 0 & 1 \\ \frac{1}{2} & \frac{1}{2} & 0 \end{bmatrix}.$$

It follows that $P^{-1}AP$ is the required diagonal matrix.

$$P^{-1}AP = \begin{bmatrix} -\frac{1}{2} & \frac{1}{2} & 0 \\ 0 & 0 & 1 \\ \frac{1}{2} & \frac{1}{2} & 0 \end{bmatrix} \begin{bmatrix} 3 & -2 & 0 \\ -2 & 3 & 0 \\ 0 & 0 & 5 \end{bmatrix} \begin{bmatrix} -1 & 0 & 1 \\ 1 & 0 & 1 \\ 0 & 1 & 0 \end{bmatrix}$$

$$= \begin{bmatrix} 5 & 0 & 0 \\ 0 & 5 & 0 \\ 0 & 0 & 1 \end{bmatrix}.$$

There is no preferred order for the columns of P. Since the ith diagonal entry of $P^{-1}AP$ is an eigenvalue for the ith column vector of P, changing the order of the columns of P merely changes the order of the eigenvalues on the diagonal of $P^{-1}AP$.

Thus, had we written

$$P = \begin{bmatrix} -1 & 1 & 0 \\ 1 & 1 & 0 \\ 0 & 0 & 1 \end{bmatrix}$$

in the last example, we would have obtained

$$P^{-1}AP = \begin{bmatrix} 5 & 0 & 0 \\ 0 & 1 & 0 \\ 0 & 0 & 5 \end{bmatrix}.$$

 Show that the matrix A is diagonalizable where:

$$A = \begin{bmatrix} 0 & 0 & 0 \\ 0 & 1 & 0 \\ 1 & 0 & 1 \end{bmatrix}.$$

 The characteristic equation of A is $\det(\lambda I - A) = 0$. Thus,

$$\det \begin{vmatrix} \lambda & 0 & 0 \\ 0 & \lambda - 1 & 0 \\ -1 & 0 & \lambda - 1 \end{vmatrix} = 0,$$

or $\lambda(\lambda - 1)(\lambda - 1) = 0$. Then the eigenvalues are $\lambda_1 = 0$, $\lambda_2 = 1$, $\lambda_3 = 1$. Now, to find the eigenvectors corresponding to $\lambda = 1$, solve the equation $(1I - A)\vec{x} = 0$ for $\vec{x}$:

$$\begin{bmatrix} 1 & 0 & 0 \\ 0 & 0 & 0 \\ -1 & 0 & 0 \end{bmatrix} \begin{bmatrix} x_1 \\ x_2 \\ x_3 \end{bmatrix} = \begin{bmatrix} 0 \\ 0 \\ 0 \end{bmatrix}.$$

Solving the above system yields $x_1 = 0$, $x_2 = r$, $x_3 = s$, where r and s are any real numbers. Thus,

$$x = \begin{bmatrix} 0 \\ r \\ s \end{bmatrix} = \begin{bmatrix} 0 \\ r \\ 0 \end{bmatrix} + \begin{bmatrix} 0 \\ 0 \\ s \end{bmatrix} = r \begin{bmatrix} 0 \\ 1 \\ 0 \end{bmatrix} + s \begin{bmatrix} 0 \\ 0 \\ 1 \end{bmatrix}$$

Thus, $\vec{v}_2$ and $\vec{v}_3$ are $\begin{bmatrix} 0 \\ 1 \\ 0 \end{bmatrix}$ and $\begin{bmatrix} 0 \\ 0 \\ 1 \end{bmatrix}$, two linearly independent eigen-

vectors associated with $\lambda = 1$. Now look for an eigenvector associated with $\lambda_1 = 0$. We have to solve $(0I - A)\,\vec{x} = 0$. Because $0I$ is the zero matrix, we have only $-A\,\vec{x} = 0$, or

$$\begin{bmatrix} 0 & 0 & 0 \\ 0 & -1 & 0 \\ -1 & 0 & -1 \end{bmatrix} \begin{bmatrix} x_1 \\ x_2 \\ x_3 \end{bmatrix} = \begin{bmatrix} 0 \\ 0 \\ 0 \end{bmatrix}.$$

A solution is any vector of the form $\begin{bmatrix} t \\ 0 \\ -t \end{bmatrix}$ for any real number t.

Thus, $x_1 = \begin{bmatrix} 1 \\ 0 \\ -1 \end{bmatrix}$ is an eigenvector associated with $\lambda_1 = 0$. Since $\vec{x}_1$,

$\vec{x}_2$, and $\vec{x}_3$ are linearly independent, A can be diagonalized. Note that an $n \times n$ matrix may fail to be diagonalizable either because all the roots of its characteristic polynomial are not real numbers or because it does not have n linearly independent eigenvectors.

 Diagonalize the matrix

$$A = \begin{bmatrix} 1 & 2 \\ -1 & 4 \end{bmatrix}$$

and find a diagonalizer for it.

If A is diagonalizable, there exists a matrix P such that PAP^{-1} is a diagonal matrix. P is called the diagonalizer of A. The diagonal elements of PAP^{-1} are the eigenvalues of A. The characteristic equation of A is $\det(\lambda I - A) = 0$. Then

$$\det \begin{vmatrix} \lambda - 1 & -2 \\ 1 & \lambda - 4 \end{vmatrix} = 0,$$

or

$$(\lambda - 1)(\lambda - 4) + 2 = 0,$$

or

$$\lambda^2 - 5\lambda + 6 = (\lambda - 3)(\lambda - 2) = 0.$$

It follows that the eigenvalues of A are $\lambda_1 = 2$ and $\lambda_2 = 3$.

To find eigenvectors for A, form the vector $\begin{bmatrix} x_1 \\ x_2 \end{bmatrix}$ and set

$$(\lambda I - A)\begin{bmatrix} x_1 \\ x_2 \end{bmatrix} = 0 \text{ for } \lambda = 2, 3.$$

For $\lambda = 2$,

$$\begin{bmatrix} 1 & -2 \\ 1 & -2 \end{bmatrix}\begin{bmatrix} x_1 \\ x_2 \end{bmatrix} = \begin{bmatrix} 0 \\ 0 \end{bmatrix},$$

or $x_1 - 2x_2 = 0$. Thus, $x_1 = \begin{bmatrix} 2 \\ 1 \end{bmatrix}$ is an eigenvector corresponding to $\lambda = 2$.

Similarly, for $\lambda = 3$, we find $\begin{bmatrix} 1 \\ 1 \end{bmatrix}$ is an eigenvector.

Therefore,

$$P = \begin{bmatrix} 2 & 1 \\ 1 & 1 \end{bmatrix} \text{ diagonalizes } A = \begin{bmatrix} 1 & 2 \\ -1 & 4 \end{bmatrix},$$

yielding the diagonal form

$$D = \begin{bmatrix} 2 & 0 \\ 0 & 3 \end{bmatrix}.$$

Verify the following rules by giving examples:

a) If A is an $n \times n$ diagonal matrix and B is an $n \times n$ matrix, each row of AB is then just the product of the diagonal entry of A times the corresponding row of B.

b) If B is a diagonal matrix, each column of AB is just the product of the corresponding column of A with the corresponding diagonal entry of B.

 a) Let

$$A = \begin{bmatrix} 2 & 0 & 0 \\ 0 & -1 & 0 \\ 0 & 0 & 3 \end{bmatrix} \text{ and } B = \begin{bmatrix} 4 & 2 & 1 \\ -1 & 0 & 6 \\ 2 & 1 & -3 \end{bmatrix}.$$

Then

$$AB = \begin{bmatrix} 2 & 0 & 0 \\ 0 & -1 & 0 \\ 0 & 0 & 3 \end{bmatrix} \begin{bmatrix} 4 & 2 & 1 \\ -1 & 0 & 6 \\ 2 & 1 & -3 \end{bmatrix}$$

$$= \begin{bmatrix} 8+0+0 & 4+0+0 & 2+0+0 \\ 0+1+0 & 0+0+0 & 0-6+0 \\ 0+0+6 & 0+0+3 & 0+0+9 \end{bmatrix} = \begin{bmatrix} 8 & 4 & 2 \\ 1 & 0 & -6 \\ 6 & 3 & -9 \end{bmatrix}.$$

This shows that each row of AB is the product of the diagonal element of A and the corresponding row of B.

b) Let

$$A = \begin{bmatrix} 3 & 0 & -1 & 2 \\ -1 & 1 & 0 & 1 \\ 4 & -1 & -2 & 1 \\ 0 & 1 & 3 & -4 \end{bmatrix} \text{ and } B = \begin{bmatrix} 4 & 0 & 0 & 0 \\ 0 & 0 & 0 & 0 \\ 0 & 0 & 3 & 0 \\ 0 & 0 & 0 & -2 \end{bmatrix}.$$

Then

$$AB = \begin{bmatrix} 3 & 0 & -1 & 2 \\ -1 & 1 & 0 & 1 \\ 4 & -1 & -2 & 1 \\ 0 & 1 & 3 & -4 \end{bmatrix} \begin{bmatrix} 4 & 0 & 0 & 0 \\ 0 & 0 & 0 & 0 \\ 0 & 0 & 3 & 0 \\ 0 & 0 & 0 & -2 \end{bmatrix}$$

$$= \begin{bmatrix} 12 & 0 & -3 & -4 \\ -4 & 0 & 0 & -2 \\ 16 & 0 & -6 & -2 \\ 0 & 0 & 9 & 8 \end{bmatrix}$$

Now use this rule to take the powers of a diagonal matrix. Find A^2 where A is the first matrix above.

$$\begin{bmatrix} 2 & 0 & 0 \\ 0 & -1 & 0 \\ 0 & 0 & 3 \end{bmatrix}^2 = \begin{bmatrix} 2 & 0 & 0 \\ 0 & -1 & 0 \\ 0 & 0 & 3 \end{bmatrix}\begin{bmatrix} 2 & 0 & 0 \\ 0 & -1 & 0 \\ 0 & 0 & 3 \end{bmatrix}$$

$$= \begin{bmatrix} 4 & 0 & 0 \\ 0 & 1 & 0 \\ 0 & 0 & 9 \end{bmatrix}$$

First find A^3:

$$A^3 = \begin{bmatrix} 2 & 0 & 0 \\ 0 & -1 & 0 \\ 0 & 0 & 3 \end{bmatrix}^3 = \begin{bmatrix} 2 & 0 & 0 \\ 0 & -1 & 0 \\ 0 & 0 & 3 \end{bmatrix}\begin{bmatrix} 2 & 0 & 0 \\ 0 & -1 & 0 \\ 0 & 0 & 3 \end{bmatrix}^2$$

$$= \begin{bmatrix} 2 & 0 & 0 \\ 0 & -1 & 0 \\ 0 & 0 & 3 \end{bmatrix}\begin{bmatrix} 4 & 0 & 0 \\ 0 & 1 & 0 \\ 0 & 0 & 9 \end{bmatrix} = \begin{bmatrix} 8 & 0 & 0 \\ 0 & -1 & 0 \\ 0 & 0 & 27 \end{bmatrix}$$

Observe how easy it is to take the powers of a diagonal matrix. Now continue with the further powers.

$$\begin{bmatrix} 2 & 0 & 0 \\ 0 & -1 & 0 \\ 0 & 0 & 3 \end{bmatrix}^{10} = \begin{bmatrix} 2^{10} & 0 & 0 \\ 0 & 1 & 0 \\ 0 & 0 & 3^{10} \end{bmatrix} = \begin{bmatrix} 1024 & 0 & 0 \\ 0 & 1 & 0 \\ 0 & 0 & 59049 \end{bmatrix}.$$

Thus, each column of AB is the product of a column of A and the corresponding diagonal element of B.

5.3 Symmetric Matrices

A square matrix A is said to be orthogonally diagonalizable if there is an orthogonal matrix P such that $P^{-1}AP$ is diagonal. The matrix P is said to orthogonally diagonalize A.

The matrix A with the property $A = A^t$ is said to be symmetric.

EXAMPLE

$\begin{bmatrix} 1 & 2 \\ 2 & 1 \end{bmatrix}$ is a symmetric matrix.

The following is the procedure for orthogonally diagonalizing a symmetric matrix A:

a) Find a basis for each eigenspace of A.

b) Find an orthonormal basis for each eigenspace by applying the Gram-Schmidt process to each basis.

c) Form the matrix P using the orthonormal bases as column vectors; P orthogonally diagonalizes A.

THEOREM

a) The characteristic equation of a symmetric matrix A has only real roots.

b) If an eigenvalue λ of a symmetric matrix is repeated L times as a root of the characteristic equation, then the eigenspace corresponding to λ is L-dimensional.

Problem Solving Examples:

Define a symmetric matrix. Is every symmetric matrix similar to a diagonal matrix?

The transpose A^t of A is the matrix obtained from A by interchanging the rows and columns of A. A matrix A is symmetric if $A^t = A$. For example:

$A = \begin{bmatrix} 2 & 1 \\ 1 & 3 \end{bmatrix}$ is symmetric since $A^t = \begin{bmatrix} 2 & 1 \\ 1 & 3 \end{bmatrix}$,

while

$A = \begin{bmatrix} 2 & 1 \\ 2 & 3 \end{bmatrix}$ is not symmetric since $A^t = \begin{bmatrix} 2 & 2 \\ 1 & 3 \end{bmatrix} \neq A$.

If a matrix is symmetric, then it is similar to a diagonal matrix. The characteristic polynomial of a symmetric matrix has only real roots and for each root of multiplicity L, one can find L independent characteristic vectors.

If A is symmetric we can actually find an orthogonal matrix P such that $P^{-1}AP$ is diagonal. The orthogonal matrix is a matrix whose columns are orthonormal. Thus, a symmetric matrix is always similar to a diagonal matrix.

 Find an orthogonal matrix P that diagonalizes

$$A = \begin{bmatrix} 4 & 2 & 2 \\ 2 & 4 & 2 \\ 2 & 2 & 4 \end{bmatrix}.$$

A The matrix A is symmetric. Construct a matrix P whose column vectors form an orthonormal set of eigenvectors of A. This can be done as follows:

1) Find a basis for each eigenspace of A.

2) Apply the Gram-Schmidt process to each of these bases to obtain an orthonormal basis for each eigenspace.

3) Form the matrix P whose columns are the basis vectors constructed in Step 2; this matrix orthogonally diagonalizes A. The characteristic equation of A is

$$\det(\lambda I - A) = \det \begin{vmatrix} \lambda - 4 & -2 & -2 \\ -2 & \lambda - 4 & -2 \\ -2 & -2 & \lambda - 4 \end{vmatrix} = 0,$$

or

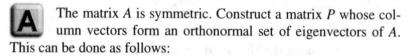

$$(\lambda - 4)\left[(\lambda - 4)^2 - 4\right] - (-2)\left[-2(\lambda - 4) - 4\right] + (-2)\left[4 + 2(\lambda - 4)\right] = 0$$
$$(\lambda - 4)(\lambda - 6)(\lambda - 2) - 4(\lambda - 2) - 4(\lambda - 2) = 0$$
$$(\lambda - 2)\left[\lambda^2 - 10\lambda + 24 - 4 - 4\right] = (\lambda - 2)^2(\lambda - 8) = 0$$

Thus, the eigenvalues of A are $\lambda = 2$ and $\lambda = 8$. To find the eigenvectors, solve the equation $(lI - A)\vec{x} = 0$ for $\vec{x}$. First, with $1 = 2$, $(2I - A)\vec{x} = 0$, or

$$\begin{bmatrix} -2 & -2 & -2 \\ -2 & -2 & -2 \\ -2 & -2 & -2 \end{bmatrix} \begin{bmatrix} x_1 \\ x_2 \\ x_3 \end{bmatrix} = \begin{bmatrix} 0 \\ 0 \\ 0 \end{bmatrix}.$$

Solving this system gives $x_1 + x_2 + x_3 = 0$. So,

$$\vec{x}_1 = \begin{bmatrix} -1 \\ 1 \\ 0 \end{bmatrix} \text{ and } \vec{x}_2 = \begin{bmatrix} -1 \\ 0 \\ 1 \end{bmatrix}$$

are two linearly independent vectors of this form. $\vec{x}_1$ and $\vec{x}_2$ form a basis for the eigenspace corresponding to $\lambda = 2$.

Applying the Gram-Schmidt process

$$\left| \vec{x} \right| = \sqrt{(-1)^2 + (1)^2 + 0^2} = \sqrt{2}.$$

Therefore,

$$\vec{v}_1 = \frac{\vec{x}_1}{\left| \vec{x}_1 \right|} = \frac{1}{\sqrt{2}} \begin{bmatrix} -1 \\ 1 \\ 0 \end{bmatrix} = \begin{bmatrix} -\frac{1}{\sqrt{2}} \\ \frac{1}{\sqrt{2}} \\ 0 \end{bmatrix},$$

and,

$$\vec{w}_2 = \vec{x}_2 - (\vec{x}_2, \vec{v}_1)\vec{v}_1.$$

Therefore,

$$\vec{w}_2 = \begin{bmatrix} -1 \\ 0 \\ 1 \end{bmatrix} - \frac{1}{\sqrt{2}} \begin{bmatrix} -\frac{1}{\sqrt{2}} \\ \frac{1}{\sqrt{2}} \\ 0 \end{bmatrix},$$

$$\vec{w}_2 = \begin{bmatrix} -\frac{1}{\sqrt{2}} \\ -\frac{1}{\sqrt{2}} \\ 1 \end{bmatrix},$$

and, hence,

$$\vec{v}_2 = \frac{\vec{w}_2}{\left| \vec{w}_2 \right|} = \frac{1}{\sqrt{6}} \begin{bmatrix} -1 \\ -1 \\ 2 \end{bmatrix}$$

$$= \begin{bmatrix} -\frac{1}{\sqrt{6}} \\ -\frac{1}{\sqrt{6}} \\ \frac{2}{\sqrt{6}} \end{bmatrix}.$$

Now let $\lambda = 8$. Then $(8I - A)\,\vec{x} = 0$, or

$$\begin{bmatrix} 4 & -2 & -2 \\ -2 & 4 & -2 \\ -2 & -2 & 4 \end{bmatrix} \begin{bmatrix} x_1 \\ x_2 \\ x_3 \end{bmatrix} = \begin{bmatrix} 0 \\ 0 \\ 0 \end{bmatrix}.$$

Thus,

$$\vec{x}_3 = \begin{bmatrix} 1 \\ 1 \\ 1 \end{bmatrix}$$

forms a basis for the eigenspace corresponding to $\lambda = 8$. Applying the Gram-Schmidt process to $\vec{x}_3$ yields

$$\vec{v_3} = \begin{bmatrix} \frac{1}{\sqrt{3}} \\ \frac{1}{\sqrt{3}} \\ \frac{1}{\sqrt{3}} \end{bmatrix}.$$

By construction $\langle v_1, v_2 \rangle = 0$; further, $\langle v_1, v_3 \rangle = \langle v_2, v_3 \rangle = 0$ so that $\{v_1, v_2, v_3\}$ is an orthonormal set of eigenvectors. Thus,

$$P = \begin{bmatrix} -\frac{1}{\sqrt{2}} & -\frac{1}{\sqrt{6}} & \frac{1}{\sqrt{3}} \\ \frac{1}{\sqrt{2}} & -\frac{1}{\sqrt{6}} & \frac{1}{\sqrt{3}} \\ 0 & \frac{2}{\sqrt{6}} & \frac{1}{\sqrt{3}} \end{bmatrix}$$

orthogonally diagonalizes A. Thus, P is an orthonormal set of eigenvectors and $P^{-1}AP$ is a diagonal matrix.

 Find the eigenvalues and an orthonormal basis for the eigenspace of A where,

$$A = \begin{bmatrix} 1 & 2 & 0 \\ 2 & 1 & 0 \\ 0 & 0 & 3 \end{bmatrix}.$$

A A is a symmetric matrix. An important result concerning symmetric matrices is that all eigenvalues of a symmetric matrix are real. Form the matrix

$$(\lambda I - A) = \begin{bmatrix} \lambda & 0 & 0 \\ 0 & \lambda & 0 \\ 0 & 0 & \lambda \end{bmatrix} - \begin{bmatrix} 1 & 2 & 0 \\ 2 & 1 & 0 \\ 0 & 0 & 3 \end{bmatrix}$$

$$= \begin{bmatrix} \lambda - 1 & -2 & 0 \\ -2 & \lambda - 1 & 0 \\ 0 & 0 & \lambda - 3 \end{bmatrix}$$

Now, expanding along the third column yields

$$\det(\lambda I - A) = (\lambda - 3)\begin{vmatrix} \lambda - 1 & -2 \\ -2 & \lambda - 1 \end{vmatrix}$$

$$= (\lambda - 3)(\lambda^2 - 2\lambda + 1 - 4)$$

$$= (\lambda - 3)(\lambda - 3)(\lambda + 1)$$

The characteristic equation of A is $(\lambda - 3)^2 (\lambda + 1) = 0$ and, therefore, the eigenvalues of A are $\lambda = 3$ and $\lambda = -1$.

Now find eigenvectors corresponding to $\lambda_1 = 3$. To do this, solve $(\lambda I - A)\vec{x} = 0$ for $\vec{x}$ with $\lambda = 3$.

$$(3I - A)\vec{x} = 0$$

or,

$$\begin{bmatrix} 2 & -2 & 0 \\ -2 & 2 & 0 \\ 0 & 0 & 0 \end{bmatrix}\begin{bmatrix} x_1 \\ x_2 \\ x_3 \end{bmatrix} = \begin{bmatrix} 0 \\ 0 \\ 0 \end{bmatrix}.$$

This is equivalent to

$$2x_1 - 2x_2 = 0 \quad \text{or} \quad x_1 - x_2 = 0.$$
$$-2x_1 + 2x_2 = 0$$

Solving this system gives $x_1 = s, x_2 = s, x_3 = t$. Therefore,

$$X = \begin{bmatrix} s \\ s \\ t \end{bmatrix} = \begin{bmatrix} s \\ s \\ 0 \end{bmatrix} + \begin{bmatrix} 0 \\ 0 \\ t \end{bmatrix} = s\begin{bmatrix} 1 \\ 1 \\ 0 \end{bmatrix} + \begin{bmatrix} 0 \\ 0 \\ 1 \end{bmatrix}$$

or

$$x_1 = \begin{bmatrix} 1 \\ 1 \\ 0 \end{bmatrix} \quad x_2 = \begin{bmatrix} 0 \\ 0 \\ 1 \end{bmatrix}.$$

Note that $\vec{x}_1$ and $\vec{x}_2$ are orthogonal to each other since $\langle x_1, x_2 \rangle = 0$.

Next, normalize $\vec{x}_1$ and $\vec{x}_2$ to obtain the unit orthogonal solutions by replacing x_i with

$$\frac{x_i}{|x_i|}.$$

Since $\left|\vec{x}_1\right| = \sqrt{2}$ and $\left|\vec{x}_2\right| = 1$,

$$\vec{u}_1 = \begin{bmatrix} \frac{1}{\sqrt{2}} \\ \frac{1}{\sqrt{2}} \\ 0 \end{bmatrix}; \quad \vec{u}_2 = \begin{bmatrix} 0 \\ 0 \\ 1 \end{bmatrix}$$

and they form a basis for the eigenspace corresponding to $\lambda = 3$. To find the eigenvectors corresponding to $\lambda = -1$, solve $(\lambda I - A)\vec{x} = 0$ for $\vec{x}$ with $\lambda = -1$.

$(-1I - A)\vec{x} = 0$ or,

$$\begin{bmatrix} -2 & -2 & 0 \\ -2 & -2 & 0 \\ 0 & 0 & -4 \end{bmatrix} \begin{bmatrix} x_1 \\ x_2 \\ x_3 \end{bmatrix} = 0.$$

Carrying out the indicated matrix multiplication,

$$\begin{array}{ll} -2x_1 - 2x_2 = 0 & x_1 + x_2 = 0 \\ -2x_1 - 2x_2 = 0 \quad \text{or} & x_3 = 0 \\ \quad\quad\; -4x_3 = 0 & \end{array}$$

Solving this system gives

$$\vec{x}_3 = \begin{bmatrix} 1 \\ -1 \\ 0 \end{bmatrix}.$$

Now, normalize $\vec{x}_3$ to obtain the unit orthogonal solution. Thus,

$$u_3 = \begin{bmatrix} \frac{1}{\sqrt{2}} \\ -\frac{1}{\sqrt{2}} \\ 0 \end{bmatrix}$$

forms a basis for eigenspace corresponding to $\lambda = -1$. Since $\langle u_1, u_3 \rangle = 0$ and $\langle u_2, u_3 \rangle = 0$, $\{u_1, u_2, u_3\}$ is an orthonormal basis of R^3.

In general, if A is a symmetric $n \times n$ matrix, then the eigenvectors of A contain an orthonormal basis of $\mathbf{R}^n$.

 Let

$$A = \begin{bmatrix} 2 & 1 & 1 \\ 1 & 2 & 1 \\ 1 & 1 & 2 \end{bmatrix}.$$ Find a (real) orthogonal matrix P such

that P^tAP is diagonal.

 A is a symmetric matrix. Hence, it has real eigenvalues. First, find the characteristic polynomial of A:

$$\det(I - A) = \begin{vmatrix} \lambda - 2 & -1 & -1 \\ -1 & \lambda - 2 & -1 \\ -1 & -1 & \lambda - 2 \end{vmatrix}$$

$$= (\lambda - 2)\begin{vmatrix} \lambda - 2 & -1 \\ -1 & \lambda - 2 \end{vmatrix} - (-1)\begin{vmatrix} -1 & -1 \\ -1 & \lambda - 2 \end{vmatrix} + (-1)\begin{vmatrix} -1 & \lambda - 2 \\ -1 & -1 \end{vmatrix}$$

$$= (\lambda - 2)\big[(\lambda - 2)(\lambda - 2) - 1\big] + \big[-(\lambda - 2) - 1\big] - \big[1 + (\lambda - 2)\big]$$

$$= (\lambda - 2)(\lambda - 3)(\lambda - 1) - (\lambda - 1) - (\lambda - 1)$$

$$= (\lambda - 1)\big[\lambda^2 - 5\lambda + 6 - 1 - 1\big]$$

$$= (\lambda - 1)(\lambda - 1)(\lambda - 4)$$

Therefore, the characteristic equation of A is $(\lambda - 1)^2 (\lambda - 2) = 0$, and the eigenvalues are $\lambda = 1$ and $\lambda = 4$. To obtain eigenvectors corresponding to $\lambda = 1$, solve the equation $(1I - A) \vec{x} = 0$:

$$\begin{bmatrix} -1 & -1 & -1 \\ -1 & -1 & -1 \\ -1 & -1 & -1 \end{bmatrix} \begin{bmatrix} x_1 \\ x_2 \\ x_3 \end{bmatrix} = \begin{bmatrix} 0 \\ 0 \\ 0 \end{bmatrix},$$

or $x_1 + x_2 + x_3 = 0$. Thus,

$$\vec{x_1} = \begin{bmatrix} 1 \\ -1 \\ 0 \end{bmatrix} \text{ and } \vec{x_2} = \begin{bmatrix} 1 \\ 1 \\ -2 \end{bmatrix}$$

are the eigenvectors corresponding to the eigenvalue $\lambda = 1$. For $\lambda = 4$, $(4I - A) \vec{x} = 0$, or

$$\begin{bmatrix} 2 & -1 & -1 \\ -1 & 2 & -1 \\ -1 & -1 & 2 \end{bmatrix} \begin{bmatrix} x_1 \\ x_2 \\ x_3 \end{bmatrix} = \begin{bmatrix} 0 \\ 0 \\ 0 \end{bmatrix},$$

or

$$2x_1 - x_2 - x_3 = 0$$
$$-x_1 + 2x_2 - x_3 = 0$$
$$-x_1 - x_2 + 2x_3 = 0$$

Thus,

$$\vec{x_3} = \begin{bmatrix} 1 \\ 1 \\ 1 \end{bmatrix} \text{ is an eigenvector associated with } \lambda = 4.$$

Next, normalize $\vec{x_1}$, $\vec{x_2}$, $\vec{x_3}$ to obtain the unit orthogonal solutions,

$$\vec{u_1} = \frac{\vec{x_1}}{|\vec{x_1}|}, \quad \vec{u_2} = \frac{\vec{x_2}}{|\vec{x_2}|}, \quad \vec{u_3} = \frac{\vec{x_3}}{|\vec{x_3}|}.$$

Thus,

$$\vec{u_1} = \begin{bmatrix} \frac{1}{\sqrt{2}} \\ -\frac{1}{\sqrt{2}} \\ 0 \end{bmatrix}, \quad \vec{u_2} = \begin{bmatrix} \frac{1}{\sqrt{6}} \\ \frac{1}{\sqrt{6}} \\ -\frac{2}{\sqrt{6}} \end{bmatrix}, \quad \vec{u_3} = \begin{bmatrix} \frac{1}{\sqrt{3}} \\ \frac{1}{\sqrt{3}} \\ \frac{2}{\sqrt{3}} \end{bmatrix}.$$

If P is the matrix whose columns are the $\vec{u_i}$ respectively,

$$P = \begin{bmatrix} \frac{1}{\sqrt{2}} & \frac{1}{\sqrt{6}} & \frac{1}{\sqrt{3}} \\ -\frac{1}{\sqrt{2}} & \frac{1}{\sqrt{6}} & \frac{1}{\sqrt{3}} \\ 0 & -\frac{2}{\sqrt{6}} & \frac{1}{\sqrt{3}} \end{bmatrix}$$

and

$$P^t A P = \begin{bmatrix} 1 & 0 & 0 \\ 0 & 1 & 0 \\ 0 & 0 & 4 \end{bmatrix}.$$

Quiz: Eigenvalues and Eigenvectors

1. The eigenvalue which corresponds to the eigenvector

    $\begin{bmatrix} 3 \\ 2 \end{bmatrix}$ for $M = \begin{bmatrix} 1 & -3 \\ -2 & 2 \end{bmatrix}$ is

    (A) 1.   (D) –4.

    (B) 4.   (E) 2.

    (C) –1.

2. Find k so that the matrix

    $$A = \begin{bmatrix} k & 1 & 2 \\ 1 & 2 & k \\ 1 & 2 & 3 \end{bmatrix}$$

has eigenvalue $\lambda = 1$.

(A) $\dfrac{1}{2}$.

(D) 1.

(B) $-\dfrac{1}{2}$.

(E) -1.

(C) 0.

3. The eigenvalues for the initial value-eigenvalue problem

$$y'' + \lambda y = 0$$
$$y(0) = 0; \quad y(\pi) = 0$$

are given by

(A) 1, 2, 3, 4, ...

(D) 0, ±1, ±4, ±9, ±16, ...

(B) 1, 4, 9, 16, ...

(E) ..., −3, −2, −1.

(C) 0, ±1, ±2, ±3, ±4, ...

4. Given that (1, 2, 3) is an eigenvector for the matrix $\begin{bmatrix} 2 & 3 & -1 \\ 3 & 2 & 1 \\ 2 & 2 & 3 \end{bmatrix}$,

find the corresponding eigenvalue.

(A) 5.

(B) 4.

(C) 3.

(D) 2.

(E) None of these choices is correct.

5. Which of the following is an eigenvalue of the matrix

$$A = \begin{bmatrix} 1 & 3 & 3 & 3 \\ 3 & 1 & 3 & 3 \\ 3 & 3 & 1 & 3 \\ 3 & 3 & 3 & 1 \end{bmatrix}?$$

(A) −1.   (D) 2.

(B) −2.   (E) 0.

(C) 1.

6. The eigenvalues of the matrix

$$\begin{vmatrix} 1 & -1 & 0 \\ -2 & 1 & 1 \\ 2 & 0 & 1 \end{vmatrix}$$

are as follows

(A) $1, \dfrac{\left(1 \pm \sqrt{5}\right)}{2}$.   (D) $-1, -\dfrac{\left(1 \pm \sqrt{5}\right)}{2}$.

(B) $1, -\dfrac{\left(1 \pm \sqrt{5}\right)}{2}$.   (E) $-1, \dfrac{\left(1 \pm \sqrt{5}\right)}{2}$.

(C) $\pm 1, \dfrac{\left(1 \pm \sqrt{5}\right)}{2}$.

7. Let

$$A = \begin{bmatrix} 1 & 0 & 0 & 0 \\ -2 & 1 & 0 & 0 \\ -2 & -2 & 1 & 0 \\ -4 & -3 & -2 & 2 \end{bmatrix}$$

be a 3×3 matrix viewed as a linear transformation from R^4 to R^4. What is the dimension of the eigenspace corresponding to the eigenvalue $\lambda = 1$?

(A) 4.

(D) 1.

(B) 3.

(E) 0.

(C) 2.

8. A real, symmetric matrix is called positive definite if $\vec{x}^t A \vec{x} > 0$ for all x in R^n. Which of the following is positive definite?

(A) $\begin{bmatrix} 4 & 2 \\ 2 & 1 \end{bmatrix}$.

(D) $\begin{bmatrix} 0 & 1 \\ 1 & 0 \end{bmatrix}$.

(B) $\begin{bmatrix} 0 & 1 \\ 1 & -1 \end{bmatrix}$.

(E) $\begin{bmatrix} 1 & -2 \\ -2 & 0 \end{bmatrix}$.

(C) $\begin{bmatrix} 2 & 0 \\ 0 & -2 \end{bmatrix}$.

9. Let A and B by $n \times n$ symmetric matrices. Which of the following is a necessary and sufficient condition for AB to be symmetric?

(A) BA is skew-symmetric.

(B) A and B are nonsingular.

(C) $|AB| = |BA|$.

(D) A and B commute.

(E) B is Hermitian.

10. Let $A = \begin{bmatrix} 2 & 1 \\ -1 & x \end{bmatrix}$. For what values of x does A possess a repeated eigenvalue?

    (A) $\{0, 4\}$.   (D) $\{1, 3\}$.

    (B) $\{0, 3\}$.   (E) $\{1, 2\}$.

    (C) $\{1, 4\}$.

ANSWER KEY

1.	(C)		6.	(B)
2.	(A)		7.	(D)
3.	(B)		8.	(A)
4.	(A)		9.	(D)
5.	(B)		10.	(A)

The Authoritative Guide...

CAREERS

for the Year 2000 and Beyond

NEWLY REVISED & EXPANDED

Everything You Need to Know to Find the Right Career

Over 250 Careers are Described in Detail:

- ➤ Earnings
- ➤ Job Descriptions
- ➤ Required Education & Training
- ➤ Working Conditions
- ➤ Advancement Opportunities
- ➤ Résumés, Cover Letters, Interviews
- ➤ The Market for College Graduates
- ➤ Tomorrow's Jobs and Outlook for the Future

REA *Research & Education Association*

Available at your local bookstore or order directly from us by sending in coupon below.

RESEARCH & EDUCATION ASSOCIATION
61 Ethel Road W., Piscataway, New Jersey 08854
Phone: (732) 819-8880   website: **www.rea.com**

VISA   **MasterCard**

Charge Card Number

☐ Payment enclosed
☐ Visa ☐ MasterCard

Expiration Date: _____ / _____
                    Mo   Yr

Please ship **"Careers for the Year 2000 and Beyond"** @ $21.95 plus $4.00 for shipping.

Name _____

Address _____

City _____ State _____ Zip _____

REA's Test Preps
The Best in Test Preparation

- REA "Test Preps" are **far more** comprehensive than any other test preparation series
- Each book contains up to **eight** full-length practice tests based on the most recent exams
- **Every** type of question likely to be given on the exams is included
- Answers are accompanied by **full** and **detailed** explanations

REA has published over 60 Test Preparation volumes in several series. They include:

Advanced Placement Exams (APs)
Biology
Calculus AB & Calculus BC
Chemistry
Computer Science
English Language & Composition
English Literature & Composition
European History
Government & Politics
Physics
Psychology
Statistics
Spanish Language
United States History

College-Level Examination Program (CLEP)
Analyzing and Interpreting Literature
College Algebra
Freshman College Composition
General Examinations
General Examinations Review
History of the United States I
Human Growth and Development
Introductory Sociology
Principles of Marketing
Spanish

SAT II: Subject Tests
American History
Biology E/M
Chemistry
English Language Proficiency Test
French
German

SAT II: Subject Tests (cont'd)
Literature
Mathematics Level IC, IIC
Physics
Spanish
Writing

Graduate Record Exams (GREs)
Biology
Chemistry
Computer Science
Economics
Engineering
General
History
Literature in English
Mathematics
Physics
Psychology
Sociology

ACT - ACT Assessment

ASVAB - Armed Services Vocational Aptitude Battery

CBEST - California Basic Educational Skills Test

CDL - Commercial Driver License Exam

CLAST - College-Level Academic Skills Test

ELM - Entry Level Mathematics

ExCET - Exam for the Certification of Educators in Texas

FE (EIT) - Fundamentals of Engineering Exam

FE Review - Fundamentals of Engineering Review

GED - High School Equivalency Diploma Exam (U.S. & Canadian editions)

GMAT - Graduate Management Admission Test

LSAT - Law School Admission Test

MAT - Miller Analogies Test

MCAT - Medical College Admission Test

MSAT - Multiple Subjects Assessment for Teachers

NJ HSPT- New Jersey High School Proficiency Test

PPST - Pre-Professional Skills Tests

PRAXIS II/NTE - Core Battery

PSAT - Preliminary Scholastic Assessment Test

SAT I - Reasoning Test

SAT I - Quick Study & Review

TASP - Texas Academic Skills Program

TOEFL - Test of English as a Foreign Language

TOEIC - Test of English for International Communication

RESEARCH & EDUCATION ASSOCIATION
61 Ethel Road W. • Piscataway, New Jersey 08854
Phone: (732) 819-8880   **website: www.rea.com**

Please send me more information about your Test Prep books

Name _____

Address _____

City _____ State _____ Zip _____